A History of the Twentieth Century in 100 Maps

A History of the Twentieth Century in 100 Maps

Tim Bryars and Tom Harper

THE UNIVERSITY OF CHICAGO PRESS
CHICAGO

Tim **Bryars** is an antiquarian map and book dealer living in London.
Tom **Harper** is the British Library's curator of antiquarian mapping and coauthor of
Magnificent Maps: Power, Propaganda and Art, published by the British Library.

The University of Chicago Press, Chicago 60637
The University of Chicago Press, Ltd., London
Text © 2014 by Tim Bryars and Tom Harper
Illustrations © 2014 by The British Library Board and other named copyright holders
All rights reserved. Published 2014.
Printed in China

23 22 21 20   3 4 5

ISBN-13: 978-0-226-20247-1 (cloth)
ISBN-13: 978-0-226-20250-1 (e-book)
DOI: 10.7208/chicago/9780226202501.001.0001

Library of Congress Cataloging-in-Publication Data

Harper, Tom, 1978– author.
  A History of the Twentieth Century in 100 maps / Tom Harper and Tim Bryars.
     pages : maps ; cm
  Includes bibliographical references and index.
  ISBN 978-0-226-20247-1 (cloth : alk. paper) — ISBN 978-0-226-20250-1 (e-book) 1. Great
Britain—History—20th century—Maps. 2. Great Britain—Colonies—History—20th century—Maps.
3. Cartography—Great Britain—History—20th century. 4. Cartography—History—20th century.
I. Bryars, Tim, author. II. Title. III. Title: History of the Twentieth Century in one hundred maps.
  GA793.7.A1H37 2014
  912.41—DC23
                            2014010722

♾ This paper meets the requirements of ANSI/NISO Z39.48-1992 (Permanence of Paper).

Acknowledgments
We would like to thank the following friends and colleagues for their insights and tireless support
at every stage of this book's creation: Peter Barber, Pinda Bryars, Kenneth Fuller, Ray Mclean,
Angus O'Neill, Drummond Walter Richards, Peter Stuchlik, Olaf Thuestad, and Laurence Worms.
Ashley Baynton-Williams, Christopher Board, Jim Caruth, John Davies, Catherine Delano-Smith,
Adrian Edwards, Philip Hatfield, Mike Heffernan, Francis Herbert, Elizabeth Van Heyningen,
Muneaki Hirano, Crispin Jewitt, John Overeem and Andy Simons.

Our thanks go out to colleagues and associates at the British Library, particularly David Way of
British Library Publishing and our editors Robert Davies and Lara Speicher for bringing the book
to life; picture editor Sally Nicholls; Chris Lee and the imaging team; designer Maggi Smith;
Mick Bucknall, Janet Grover and the retrieval team for the finding and replacing of the thousands
of maps used in researching this book, and the staff of the famous Maps Reading Room including
Nicola Beech, Gareth Burfoot, Carlos Corbin, Jo Dansie, Carlos Garcia and James Greenaway.

Special thanks are due to Ron Jones, Stan McCaffrey, Miran Norderland and Jim Sharp for making
maps; and Stephen Humphries and William Wells for using them; to collectors Ronald Griffiths
and Winfrid de Munck and to all the collectors, researchers and enthusiasts who help keep maps –
especially paper ones – very much alive.

Finally, we wish to convey personal thanks to Louise Bryan, Paula Bryars, Panagiotis Chantziaras,
Geoff Harper and Jonathan Potter for inspiring in us a lifelong love of maps.

Frontispiece: Detail of pp. 92–93
Pages 6–7: Detail of pp. 58–59

Contents

House of the Sorcerers.

Here do the Sidhe make the Water of Life

Cockpaidle Cape

The Sea of

Elfin Sound.

The Little Mermaid.

Here are Leprechauns.

Ferlie Firth.

This is Ole Luk-Oie.

Here groweth the Sacred Vervain

Here is Oberon's Palace.

Here are Neckans.

TomTitTot lives here.

Golden Strand

Kelpie Bay.

Here dwell Nixies and Water Sprites.

Little Tuck.

Troll Town

The Kelpies Hamlet

Elfin Citie.

Brownies' Huts.

This Greate Walle was builded of Stars by manie Elfin Emperours in Days Remote.

the s' Caves

Here are Snow White and Rose Red.

This is TomTitTot.

Kilmeny's House.

The Well of Youth

King of Toole's Palace.

The Marquis of Carabas His House.

Here Deirdre dwellet

Peter Piper

This is Mrs Bond

Puss in Boots

adine Bay

Green Harbour

Here King Thrushbeard dwelleth

and

eefs

Pixie Town

The Emeralde Beaches.

The Bay of

us.

A wilde Swan.

Point.

Peter Pan's House.

Silverbell Light

EXCAL

Here liveth Morgan Le Fay

Here is the Never-Never Land.

St Brandan's Isle.

Honeymouth Cove.

Here is Simple Simon.

Elfrain Cove.

The Taylor

Tom Thumb is somewhere here but he is too small to draw.

Jack's House.

Little Boy Blue.

Red Riding Hood's House.

Contrary Mary's Home.

Old Wother Hubbard's Home

Tom Piper's Home

Miss Wuffet's House.

Here Jack and Jill are married

The Palace of Old King Cole

Taffy's House.

ELFLAND

WORLD

Banbury Cross

M. Dames' Cottage.

Knave of Hearts Cottage.

This is Actæon

Humpty Dumpty.

Goosey Goosey Gander.

Jack Sprat's Cottage.

Oberon Cross.

Here Cinderella liveth with her Prince

The River of White Nymphs.

The Elfin Temple.

Here is the Sibyl.

Here are Quicksands.

This Great Gate is built of Ivory.

Introduction

THE TWENTIETH CENTURY was a golden age of map-making, an era of cartographic boom. It was the first period of near-universal map literacy, when maps proliferated and permeated almost every aspect of daily life. It was a century overshadowed by war, yet marked by tremendous social and technological change to which the examples in this book are contemporary witnesses. We have selected one hundred maps from the millions that were printed, drawn or otherwise constructed during the twentieth century; by placing these documents in their original contexts we have created a subjective but wide-ranging narrative of some of the key events and developments of the century.

The rich variety of mapping is one of the themes of this book. On the following pages are maps that were printed on handkerchiefs and on the endpapers of books; maps that were used in advertising or propaganda (fig. 1), or that illustrated government reports; maps that were strictly official and those that were entirely commercial; maps that were printed by the thousand, and highly specialist maps issued in editions of just a few dozen; maps that were envisaged as permanent keepsakes of major events, and maps that were relevant for a matter of hours or days.

Cultural and social history told through artefacts has been a particular enthusiasm of the past few decades. Maps are ideal narrators because their breadth of coverage enables them to illustrate key events in twentieth-century history, such as the outbreak of the Great War of 1914–18 and the 1969 moon landing. Their meanings and messages (which are sometimes incidental to their original purpose) shed light on broader social attitudes – towards immigration, towards women in the workplace and towards sexuality, for example.

A few maps have already gained recognition for their historical significance. The 1973 map of the world on the equal area projection devised by Arno Peters (see p. 193), or the 1933 diagram of the London Underground transport system by Harry Beck (fig. 2), have become well-known cartographic superstars, but the majority of maps on these pages have rarely, if ever, been seen in print since they first appeared.

The narrative of the book flows around the idea of a 'British' twentieth century. While the meaning of the term is nuanced and continues to evolve, we have interpreted 'British' in the broadest possible way, culturally and geographically. What emerges is a shared history that encompasses more than just the British Isles. J. R. R. Tolkien's 1954 novel *The Lord of the Rings*, which has been translated into dozens of languages including Faroese and Arabic, contains fantasy maps that have been pored over by millions worldwide (see p. 132). British planners have informed the urban landscape in Hong Kong and New Delhi as well as Glasgow. British policymakers were inextricably bound up in the Partition of India (p. 118) and the creation of Israel (p. 79), and the consequences of their decisions have been felt on a global scale, not least by naturalised Britons and their descendants from the former Empire and Commonwealth.

The British view of the world – and the rest of the world's view of the British – has also been shaped by personal as well as political activity, by servicemen on leave in wartime New York, for example (see p. 104), or backpackers in 1990s Australia. Twentieth-century Britain reacted to outside forces: to Jewish refugees fleeing the pogroms in Eastern Europe; to American tourists; to the rise and fall of fascism. A full understanding of a map of London, with its foreign references from East India Docks and Waterloo Station to Swiss Cottage and Maida Vale, demands an awareness of the rest of the world.[1] Although this book contains a view of the twentieth century through British eyes, a similar template would work with any map-making nation and culture.

Fig. 1
'Le Partage de L'Allemagne …', 1917
Private collection

Simplified maps lend themselves to propaganda and postcards for similar reasons: they provide striking, instantly recognisable images, even when reproduced in a small space. This First World War German propaganda postcard combines both elements, and it follows another golden rule of successful propaganda: what the enemy are actually saying about you may be far more inflammatory than anything you can invent. Here the cover of a French book of 1913 is reproduced with little alteration, other than a comment that the partition of Germany was planned by the French, British and Russians even before the outbreak of war. It did not happen in 1918, but it is curiously prescient of 1945.

Maps are children of their times Delano-Smith and Kain[2]

Reality in our century is not something to be faced Graham Greene[3]

Fig. 2
Diagram of the London
Underground Railway system,
H. C. Beck, 1933
London Transport Museum

Harry Beck's schematic diagram
of the London Underground,
an innovative and influential
example of industrial design and
Art Deco, was utilised by rapid
transport systems across the world.
This iconic image has featured on
stamps and posters, on t-shirts
and tea towels, and it has inspired
a host of artworks and parodies.

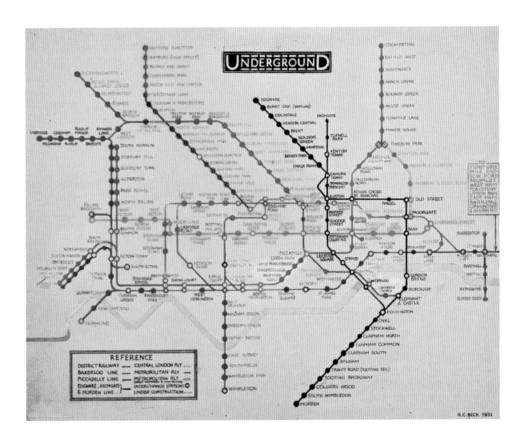

That is the primary viewpoint of the book. Its organisation is chronological rather than thematic. Maps are often discussed in highly specific contexts – various categories of transport maps, for example. This is essential to trace the development of a particular form of mapping, but limited in terms of what it can contribute to a broader understanding of the society that created the maps. We have retained conventional subdivisions of the century into eras shaped by key events such as the two world wars. However, the juxtaposition of maps that never 'belonged' together, but that were created at the same moment in history, can be revealing: E. H. Shepard's 'Hundred Acre Wood' was produced in the same year as the plan for the suppression of the general strike (1926). Our intention is to create a sense of what it was like to live through each part of the century.

One deceptively simple question that has regularly been asked over the past twenty-five years in books and articles is 'what is a map?'[4] As our selection shows, the definition can be wide, but perhaps the most interesting point about the question is that it only started to be asked in the twentieth century.[5] The mass emergence of the map obsessive is a recent phenomenon. People have not always been mesmerised by squiggly lines and symbols – they have not always had the opportunity. For much of the medieval period, and despite the introduction of printing in Europe in the fifteenth century, mapping remained not only elitist and unintelligible to the majority of people, but a highly unusual activity.[6] However, by 1900 Britain enjoyed an extremely high literacy rate, just as the language of maps gained mass appeal.

Maps and mapping have always been at the forefront of technical and technological developments in image production, reflecting an increasingly map-literate, mobile and affluent society. In early modern Europe and substantially earlier in the Far East, maps were used almost exclusively for administration, warfare and state propaganda, at an elite level. During the nineteenth century, cheaper printing costs and mass educational movements were the technical and intellectual impetus for bringing maps to a wider audience. Maps and map education were either privately sponsored by bodies such as the Society for the Diffusion of Useful Knowledge (established in 1826), or backed by the state, in the board schools set up as a consequence of the 1870 Elementary Education Act. However, warfare and imperial expansion were the principal catalysts for increased map production leading into the twentieth century. 'Theatre of war' maps, for example, satisfied the desire of an ever more map-conscious British public to track major international conflicts, such as the Crimean War of 1854–6 and Queen Victoria's 'small wars', fought in far-flung corners of the British Empire in the second half of the century.

The eighteenth century made the gentry map-literate and the nineteenth century brought map-literacy to the middle classes, but the Great War of 1914–18 extended map-mindedness to a majority of the population.[7] Hundreds of thousands volunteered for service in South Africa in 1900, but in the First World War millions of people were sent abroad or beyond their immediate surroundings for the first time in their lives. Significant portions of the populations of Belfast, Sydney or Vancouver found themselves in places such as Turkey, Belgium and Mesopotamia: 'I have seen foreign people, countries and continents, and all this only due to the war' wrote a soldier after 1918.[8]

The sheer effort, money and resources poured into the waging of the Great War drove an improvement in the production of maps as military tools, and honed the skills needed to understand and use them. This new map-mindedness did not evaporate with the 1918 Armistice: it blossomed in education, in travel and for leisure. In Britain during the 1920s, tourist maps and paid holidays went hand-in-hand, and in North America leisure time was similarly furnished with mapping.[9] Demobilised servicemen who had fought hundreds or thousands of miles away from home brought back their knowledge and their map skills.

The real key to understanding maps and their increasing usage across the twentieth century is an appreciation of their widened audience. They continued to function in the Cabinet Office and War Room, as they had since the reign of Henry VIII, accompanying peace treaties in which whole nations were created and divided by means of the lines on a map. But during the century they also became some of the most 'everyday' and ephemeral objects we possess, churned out in vast numbers by pen, printer or photocopier, or even viewed on a computer screen to assist such familiar pursuits as way-finding, learning and advertising.

It would be a mistake to discuss maps as isolated images, however. They became more prevalent just as society in general became increasingly visually minded. Increasing demand for the printed image was met by increasingly efficient means of production. Maps are distinctive types of images containing texts, differing from other pictures in that they require a key to the symbols, codes and conventions out of which they are constructed.[10] But they are produced by the same methods as other images, and these production methods changed over the twentieth century.

By the end of the nineteenth century cheaper printing costs ensured that maps could be printed, sold and distributed in quantities that were previously unimaginable. This was due, in part, to advances in paper manufacture and the mechanisation of presses, but more particularly to the development of commercially viable colour-printing techniques and variations on the lithographic

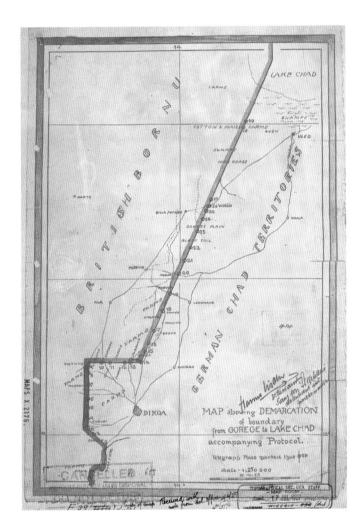

Fig. 3
Map showing demarcation of boundary from Gorege to Lake Chad, accompanying protocol. War Office, 1903. British Library

This War Office map of parts of modern-day Cameroon and Nigeria illustrates which features were of interest to the British administration, including roads, farms and natural obstacles. It also illustrates a fundamental tension between the basic concept of political boundaries which was familiar to the European map-makers, and the arbitrary nature of such boundaries given the mobile and impermanent way of life of the people who already lived there.

printing process. From around 1910 photography began to replace traditional data-gathering, such as ground survey, in map production. But the most striking changes arrived with digital technology in the 1970s. Maps no longer needed to be on paper: they could exist far more easily on a screen. 'Geographical Information Systems', satellite positioning and imaging, virtual simulation programmes and (from the 1990s) online mapping all reflected the increasing move towards web publishing.[11]

This description of the evolution of mapping should not, however, imply that map production was always being improved and refined. People have continued to draw maps by hand, and it could be argued that some digital maps are far 'worse' in quality than many maps of the hand-press period, engraved two centuries ago or more. Many modern narratives of advancement suggest that technological progress is somehow independent of its creators and the society it serves. On the contrary, maps are fundamentally connected with their makers and their users. In a sense that is the real subject of this book. A look at the broader categories of the twentieth-century map archive provides an introduction to these people.

Official maps

The most familiar, possibly notorious, type of map is the official map, produced 'in-house' for the state and its military by agencies such as the Ordnance Survey of Great Britain (OS, founded 1791 but with its origins just after the Jacobite rebellions in 1747), the United States Hydrographic Survey (1866), or the Survey of India (1767). This mapping served the full range of official and governmental business, and, although often pressed into administrative and commercial service, its origin was invariably military. These are the maps to which we attach words such as 'covert' and 'secret', since the information stored in them could be extremely valuable to an enemy and very damaging if lost. In this category we place invasion plans, trench maps from the Great War, and large multi-sheet topographical maps which, more than anything else, illustrate the relationship between the process of mapping and that of state and imperial control (fig. 3).

Of course, once maps and mapping data exist, they can be put to a variety of uses. Official maps have been used for post-war settlements and partition – a twentieth-century solution to granting a particular community some measure of autonomy or independence, such as at Versailles in 1919, and Pakistan and Israel (both 1947; see p. 118). Reconstruction and civil engineering projects, such as motorway construction or flood defences, used maps in their planning, publicity and building stages. Taxation and changes to administrative boundaries required maps to identify who was in and who was out. These maps are consistent and powerful in what they enabled institutions to achieve.

Insider maps

Mapping for the commercial sector – highly specialised maps for business – is hardly less secretive and specialised. Business and government have often gone hand-in-hand. Maps were produced to support some of the great commercial operations of the century, from construction to mineral extraction. The map data was often licensed from official sources and the intended audience was narrow. Print runs – if they ever got as far as print – were small, and their contents were often jealously guarded.

The extraction, supply and consumption of oil, for example, has shaped our daily lives as well as the course of international politics, and the process relies on maps at every stage: from prospecting to promotion and ongoing administration. Such maps could be purchased from mapping agencies such as the OS and others such as the Royal Geological Survey, and tailored to specific needs.

Occasionally, however, independent trade associations and magazines can provide a way into these specialised trades. However recondite the map made for the oil magnate may be, if it is decipherable it will reflect where his or her gaze is focused. Maps of business markets, whether they be superstores or the lines of communication linking them, visualise the geographies in order to invite exploitation.

Popular maps

To many the first image conjured by the name Ordnance Survey will not be the land survey extract for the government report, or historic mineral planning data, but the fell walker with the folded-up 'Explorer' map. This is the popular map for everyday use, the leisure map, and in the century before online map interfaces or in-car navigation systems it was the best map available to us.

Popular mapping is the term we use to describe every form of mapping produced and used outside official circles. These are the maps, atlases, globes and other cartographic objects with which we are most familiar, tailored as they are to assist us in our education, navigation and leisure time. Large commercial publishers, such as HarperCollins, Imray, Rand McNally or Bartholomew, trusted for their affiliations to authoritative geographical societies, dominated large chunks of the map trade throughout the century.

But smaller independent map companies also thrived during the twentieth century for the same reasons: decreasing production costs and the identification of specialist markets. The Geographers' A–Z Map Company (Pearsall's London map was produced in 1934) is one of the best examples, while the Philip's Planisphere (fig. 4) is possibly the most ingenious map ever produced. And maps did not need to be the preserve of the specialist. Literary maps adorn the endpapers of countless detective, historical and fantasy novels. Artists consciously employed the language and conventions of maps in their work, while video games incorporating maps emerged during the 1990s. The sheer variety of the map archive is beyond full description in a few pages.

Ephemeral maps

A lot of maps were produced in the twentieth century; millions, in fact, including a large number that were never meant to be saved. These are the maps scribbled on the backs of napkins or published in newspapers, which were normally disposed of after they had served their purpose. Imagine if every one of the 346 million maps produced by the Ordnance Survey over the course of the Second World War survived.[12] The twentieth-century's continuous map cull, born out of the intense yet short-term usefulness of the map, is both necessary and ongoing. Maps have always been for use, but not necessarily for retention.[13]

Fig. 4
Philip's Planisphere showing the principal stars visible for every hour in the year, 1991
British Library

Philip's ingenious celestial map-device, first produced in the late nineteenth century, is based on the astrolabe, an ancient scientific instrument developed in the Arabic world. By rotating the uppermost plastic ring, amateur astronomers could pinpoint in the sky each of the major stars and planets visible at the latitude of Northern Europe.

PHILIP'S
PLANISPHERE
FOR LATITUDE 51.5 NORTH
(Northern Europe, Northern USA and Canada)
SHOWS THE POSITIONS OF THE STARS AND CONSTELLATIONS FOR ANY HOUR IN THE YEAR

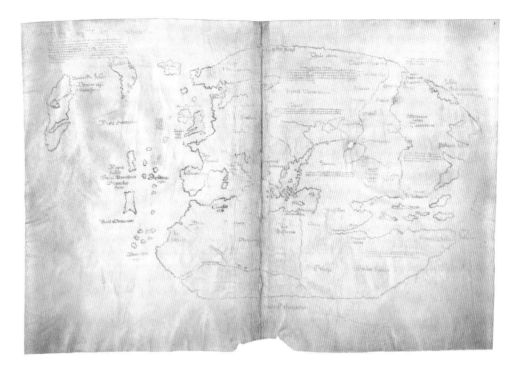

Some maps have outlived their contexts and acquired successive meanings over time. A small number of pre-twentieth-century maps are included in this book for this reason. The partly manuscript atlas of Europe produced by Gerard Mercator in 1570–2 became relevant to the twentieth century in 1979 when it was viewed as an art investment (p. 214). Another apparently old map that we have not included, the Vinland map (fig. 5), may well be a forgery with its history entirely in the twentieth century.[14] The lottery of map survival means that any story we tell using maps as the starting point will be skewed in some way. They are subject to decay, defamation and rehabilitation like the stories of any historical figure.

A century is the blinking of an eye, even in terms of human history, but in the twentieth we seem to have moved further and faster than in any other. Maps have infiltrated every part of life and society, from the map in the national newspaper to the map in the government dossier. It may seem difficult, at first glance, to spot the threads running through a period which began with the death of Queen Victoria – the very concept of a 'Queen-Empress' is now wholly alien – and that ended with the premiership of Tony Blair. The twentieth century encompasses Blériot's Type XI monoplane and NASA's Space Shuttle, and we must reconcile a world that could envisage the annihilation of mankind through conventional weapons (as recently as 1939, when MGM's pacifist cartoon 'Peace on Earth' depicted the last two men killing one another with bolt-action rifles) with the terrors of the atomic age unleashed by the Manhattan Project. At the beginning of the century, attempting to flee justice on a transatlantic liner was a rational thing to do; Dr Crippen was the first suspect to be caught as a result of wireless telegraphy in 1910. By 2000, passengers on the same transatlantic crossing were able to check their emails on board.

We will be looking at maps made to help us get from A to B (or A to Z), maps in satire or propaganda, maps in advertising, maps in education or at the forefront of scientific and technological advances, maps in literature or in games, whimsical maps made with no other purpose than to raise a smile, and maps which are rooted in some of the darkest chapters of recent human history, including total war, attempted genocide and ethnic cleansing (all terms coined during our period). Maps remind us how recent certain concepts, such as universal healthcare, really are; they can also inform us about how deeply rooted other seemingly contemporary issues are, such as oil exploitation in the Middle East.

To progress the author Graham Greene's comment a step further, maps are one of the many realities of the twentieth century; they are human-constructed filters, virtual realities laid over the real world. Of course, the 'real' world is an entirely subjective concept, and the following pages offer many different ways of defining it.

CONTROL STIRLING

END SECTION 2 343 MILES

CONTROL

22 M. 31 M.

EDINBURGH

Renfrew

CONTROL GLASGOW

Paisley

Hamilton Wishaw

Berwick

Strathaven

Galashiels

Ayr

Kelso

86 MILES

93 MILES

Moffat

Jedburgh

Dumfries

NEWCASTLE CONTROL

CONTROL CARLISLE

Durham

Penrith

Bishop Auckland

Stockton

Middlesborough

Darlington

68 MILES

Richmond

Northallerton

Scarborough

Kendal

Kirkby Lonsdale

Ripon

103 MILES

HARROGATE CONTROL York

IRISH SEA

Leeds

Selby

Hull

Blackburn

Accrington

Normanton

NORTH SEA

Bolton

Bury

Barnsley

Doncaster

CONTROL MANCHESTER

SHEFFIELD

Worksop

Lincoln

Macclesfield

Mansfield

CREWE

Burslem

Southwell

Hanley

Nottingham

182 MILES

Wymondham

Stafford

Melton Mowbray

Oakham

Norwich

Shrewsbury

Shifnal

WOLVERHAMPTON

141 MILES

Bridgnort

BIRMINGHAM

Ely

Huntingdon

Kidderminster COVENTRY

Kettering

Cambridge

Ipswich

Worcester

Wellingborough

Northampton

Hereford

Ledbury

Bedford

Colchester

Ross

Cheltenham

Buckingham

Gloucester

Luton

Swansea

Oxford

St Albans

HENDON CONTROL

END SECTION 3 TOTAL 383 MILES CONTROL

CARDIFF

BRISTOL

Bath

SECTION 1 20 MILES

LONDON

START BROOKLANDS

END SECTION 5

FINISH

Bridgewater

Tunbridge Wells

Taunton

83 MILES

SALISBURY PLAIN

CONTROL

Southampton

76 MILES

40 MILES

Horsham

CONTROL BRIGHTON

END SECTION 4 224 MILES

Dover

63 MILES

CONTROL EXETER

Honiton

Dorchester

WIGHT

Arundel

PORTSMOUTH

Weymouth

Bournemouth

Truro

ENGLISH CHANNEL

SCALE OF MILES

10 20 30 40 50

Table of Distances.

	Miles.
Brooklands to Hendon ..	20
Hendon to Harrogate ..	182
Harrogate to Newcastle ..	68
Newcastle to Edinburgh ..	93
Edinburgh to Stirling ..	31
Stirling to Glasgow.. ..	22
Glasgow to Carlisle	86
Carlisle to Manchester ..	103
Manchester to Bristol ..	141
Bristol to Exeter	65
Exeter to Salisbury Plain ..	83
Salisbury Plain to Brighton	76
Brighton to Brooklands ..	40
Total	1,010

A Weary Titan? 1900–1918

'The weary titan staggers under the too vast orb of his fate' was Colonial Secretary Joseph Chamberlain's appraisal of Britain, and Britain's role in the world, in 1902. On the face of it Britain was the world's pre-eminent power. Shades of red or pink used to denote the British Empire on maps, such as the Navy League map (p. 20), coloured a fifth of the world's surface, and in 1900 approximately a quarter of the world's people were subjects of the British Crown. Recent studies of the 'British world' equate the period that ended with the outbreak of war with 'the first wave of modern globalisation'.[15]

To an extent, however, the homogeneous red on maps of the empire is misleading. The same shade is used on self-governing dominions such as Canada and Australia; colonies with a long and involved history of contact with Britain, such as India; and newly acquired and barely explored African colonies, which often attracted a low level of investment. The balance sheet of empire is a complex issue, but overall this sprawling and diverse entity was not necessarily profitable or sustainable: Chamberlain's remark was part of an appeal to the Imperial Conference for greater support in sharing with the British taxpayer the defence and administrative 'burdens' of 'the vast empire which is yours as well as ours'.

The Navy League map is also a reminder that there were new rivals, such as Germany, for command of the oceans. Major railway schemes such as the Berlin–Baghdad railway (see p. 52) were indicators that Britain had lost her industrial lead, and that British sea power was in any case increasingly obsolete in a world where armies could be moved by rail across continents. It was an era of relative decline which opened with the international humiliation of the Boer War (see p. 18).

And yet the empire rallied to the flag in 1900 and again in 1914. The souvenir map celebrating the Relief of Ladysmith (p. 26) is a reminder of the keenness with which the British public followed the Boer War; and the comic zoomorphic map showing the European powers as scrapping dogs (p. 46) is a British contribution to the general euphoria that gripped most of the belligerent nations on the outbreak of the Great War (we have also included a Franco-Polish example on p. 48). It would be glib to suggest that the pomp and pageantry that surrounded imperial events such as the Delhi Durbar (see p. 42) should be ascribed to hubris, pure and simple. It would have been impossible for contemporaries to predict the impact of two world wars, or to assume that isolationist America would take on Britain's role as self-appointed world policeman.

In 1900 much of the world remained unexplored, or at least unmapped. This was the heroic age of polar exploration, which concluded with the safe return of Shackleton's Imperial Trans-Antarctic Expedition in 1917 (see p. 44). It was not, however, only the most desolate, inhospitable terrains that were poorly mapped: despite four centuries of European interaction with southern Africa, the maps provided to troops during the Boer War (see p. 18) were some of the worst ever issued to the British Army.

By contrast, our *Trip to the Continent* board game (p. 16) illustrates how the world was opening up for leisure travel. New forms of mass communication, such as the postcard, were also being developed: *A Map of the World, as Seen by Him* (p. 30) of 1907 is a gently humorous optical illusion – a slightly distorted map of the eastern hemisphere that becomes a woman's face. The combination of map subject and postcard medium feeds into our discussion on the growth of map literacy in this period. Other maps in this section, covering subjects such as the drug trade, the immigration debate and the relationship between Ireland and the United Kingdom, are very much of their time, but they can inform our understanding of topics that remain controversial a century later.

In 1914 Europe stumbled, or rather marched headlong with cheering and flag-waving, into a new kind of conflict for which it was wholly unprepared – and it dragged the rest of the world with it. Four years of carnage and a poisonous post-war settlement scarred nations and individuals both physically and mentally, and cast a long shadow over the rest of the century. As with all the conflict maps in this book, we have juxtaposed those that were intended to inform or influence opinion on the home front, such as *What Germany Wants!* (p. 52), with those that were used on the front line, such as our trench map of the Gallipoli peninsula (p. 50). Above all, we have attempted to convey the scale of a war that caused over 37 million casualties, brought down four empires

The route map for the 1911 Daily Mail Circuit of Britain Air Race, published in aviation news magazine *Flight* (22 July, 1911)

Only four of thirty entrants completed the course: the winner and runner-up arrived back at the starting point, Brooklands, in under twenty-four hours but the other successful

1900: *A Trip to the Continent*: Edwardian tourism

ANY RULES SHEET for this Edwardian board game vanished many years ago, but it seems straightforward enough. There are no snakes or ladders to delay or speed up the journey: no lost luggage, missed trains or banker's drafts to be collected *poste restante*. Players choose one of three routes to leave London and race to Berlin, spinning a totem or rolling dice. The map on which the game is based is so crude that one can safely discount any serious underlying educational purpose (many cartographic games of this period incorporate the capital cities of Europe, backed with useful facts and statistics about population or exports for the edification of younger players). It does, however, successfully convey where Edwardian Britons liked to go on holiday, which is in itself quite surprising. A decade later a 'race to Berlin' would have carried entirely different connotations, but in the years leading up to the Great War, Germany was one of the top tourist destinations and German culture was celebrated.

In the century between the battles of Waterloo and Mons the British 'discovered' the Continent as a direct consequence of a new mode of transport: the railway boom made travel faster, safer and cheaper at a time of rising prosperity for the middle classes. Travel would never again be the preserve of a monied elite, and by 1900 many recognisable elements of the tourist industry were in place: increased hotel capacity and other dedicated facilities in resorts, mass-produced souvenirs, group tours and travel agencies. Some Britons were already seeking out unspoiled corners where they wouldn't hear another English voice, especially a 'cockney' or lower-class one; as Shaw wrote in 1912, 'it is impossible for an Englishman to open his mouth without making some other Englishman hate or despise him' (*Pygmalion*).[16] A clear distinction was also drawn between 'travellers' and 'tourists' that isn't unfamiliar today. Among their Continental hosts, the British gained a reputation for insisting on their own cuisine wherever they went and for a lack of linguistic ability that compelled anyone connected with the hospitality industry to learn some English ('the great educator ... the missionary of the English tongue', as Jerome K. Jerome described the English-speaker who 'stands amid the strangers and jingles his gold').[17] Some of these early tourists also displayed a regrettable lack of respect for local customs. There are abundant recorded examples of (mostly Protestant) British tourists disrupting (mostly Catholic) church services in their determination to take in the sights. They were also lampooned for their curious dress – or rather, for being underdressed. Going on holiday was a chance to wear casual clothes (loud checks were particularly favoured), but dressing for the country, whether cycling in Normandy or strolling along the Avenue des Champs-Élysées, was bound to invite comment.[18]

This board game reflects turn-of-the-century travel. Entry to the Continent is via the three French Channel ports: Calais, Boulogne and Dieppe, already known for weekend breaks if not for day trips. Boulogne had also supported a thriving debtors' colony since the 1840s. The tracks unite at Amiens, passing through Belle-Époque Paris, and then cross north through Cologne, traversing a splash of blue that may represent the Rhine (with no detour for a cruise, although that had been one of the most popular pastimes a generation earlier); Berlin is the ultimate goal. Rome would have been the natural destination of the wealthy Grand Tourist of earlier times. The game also ignores Switzerland, which had been one of the biggest draws for Victorian tourists. Sherlock Holmes and Moriarty wrestle on the Reichenbach Falls because of what Conan Doyle did on his holidays (indeed, Holmes manages to enjoy a week-long walking holiday before the fatal struggle).

The popularity of Berlin as a holiday destination began to wane in the Edwardian period, as it became increasingly clear that Germany, rather than France or Russia, would be the enemy in the event of a European war. For an Englishman's view of Germany in 1900, however, we have Jerome K. Jerome's *Three Men on the Bummel*, the sequel to *Three Men on a Boat*, in which the three friends take a cycling holiday in the Black Forest. Jerome mocks neither his mildly inept middle-class tourists nor their hosts in anything but the gentlest manner. After commenting again on the German love of order, Jerome writes: 'The Germans are a good people. On the whole, perhaps the best people in the world; an amiable, unselfish, kindly people. I am positive that the vast majority of them go to heaven.'[19]

1900: 'Better than nothing at all': mapping the Boer War

THIS MAP OF the environs of Bloemfontein, capital of the Orange Free State, was published for use in the field during the second Anglo-Boer War (1899–1902). In contrast to the following map (which was printed to boost morale at home), this was for military use, despite the alarming phrase that appears in the margin: 'this map is not to be considered as absolutely accurate'. At the beginning of the twentieth century great swathes of the world remained relatively badly mapped, if mapped at all. The quality of maps available at the outset of the campaign forms a substantial topic in the Minutes of evidence taken before the Royal Commission on the War in South Africa (HMSO 1903) and they have retained their reputation of being 'amongst the worst maps ever issued to British troops'.[20] Lessons were learned, but like many others learned at great cost in the Boer War – the British Empire's first conflict in half a century against a 'European' enemy armed with modern weapons – subsequent application was patchy and haphazard. With exceptions such as the Survey of India, Britain's most comprehensive programme of overseas mapping was undertaken in the twilight of the colonial era after the Second World War (see, for example, the 1978 map of Rhodesia, p. 176).

A local South African firm, Wood and Ortlepp of Cape Town, was entrusted with compiling 'the Imperial Map of South Africa' on behalf of the Field Intelligence Department. Thirty-one sheets on a uniform scale of 1:250,000 were compiled in the nine months leading up to January 1900, which were printed between January and April. It has been estimated that of 120,000 that were printed some 95,000 were distributed. After the war ownership was to revert to John T. Wood, and examples exist which were subsequently sold to the public. The maps were bound with whatever came to hand: canvas, linen, card or paper. However, there was a certain continuity of a very personal nature: many of the maps, including our example, do not open out

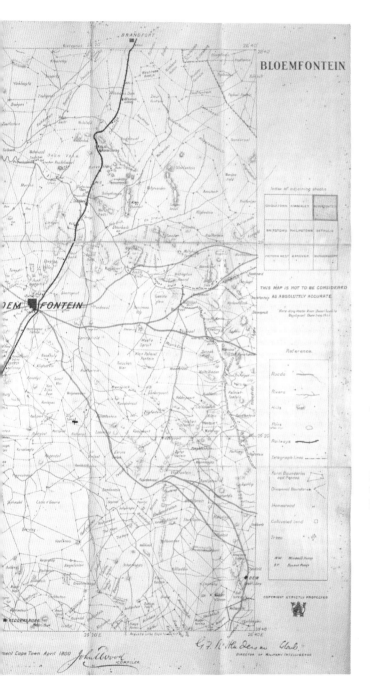

naturally to the right, suggesting that they were mostly mounted by someone left-handed.[21]

The Bloemfontein sheet was first published in February 1900 and the city was captured in March. Our example is revised to April, and the date 7/12/00 inked into the lower left margin suggests that it remained current later in the year. No attempt has been made to mark the location of either of the concentration camps established at Bloemfontein by the British. The first of these, the first camp in South Africa, opened in September 1900; within months its population outstripped that of Bloemfontein itself. It is known to have been located on a slope (to facilitate drainage) a couple of miles outside the city, and it has been suggested that it may have been near the dam marked on Bloemspruit farm, just to the east.[22]

Our map belongs to the last phase of the war. Realising that the British were winning the conventional war through sheer weight of numbers, the Boers organised Kommandos who used their superior knowledge of the terrain to mount ambushes before melting into the civilian population. Faced with a guerrilla campaign which pinned down 250,000 troops amid spiralling casualties and costs, and growing condemnation at home and abroad, the British resorted to a scorched-earth policy – controlling the terrain from a series of blockhouses, moving between them in armoured trains, destroying crops and farms, and herding the population, both African and Afrikaner, into camps.

Concentration camps were internment camps. The English term was coined during the war (although the concept was not unique, as is evident in the near contemporary *reconcentrados* of the American–Philippine War, for example). However, subsequent associations with Nazi death and labour camps (such as the camps on Alderney: see p. 106) should not distract from either the nature of the Boer camps or the tremendous loss of life that occurred in them through overcrowding, malnutrition and disease.

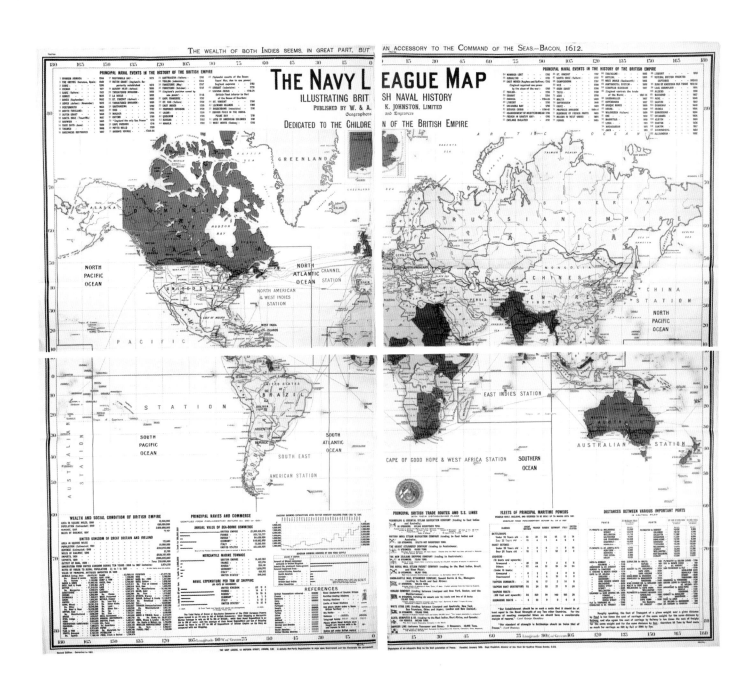

1901: A glue spread increasingly thin: the Navy League map

THE NAVY LEAGUE map is the archetypal British Empire map, a vast creation nearly 2 metres in width, published by W. & A. K. Johnston in 1899 and updated in several editions up to 1925. It shows the world on Mercator's projection with the British Isles and the empire coloured in red. Smaller strategic points, such as Gibraltar, the Cape of Good Hope, Suez, Panama and hundreds of islands, are underlined in red. Around the edge and in the spaces in between are tables and statistics recalling Britain's proud naval heritage (beginning in 1588 with the defeat of the Spanish Armada), its naval strength and its financial yield. It is 'dedicated to the children of the British Empire'.

Recent analysis of British Empire maps has dwelt upon their artificial use of dominant colour and distortion to demonstrate and uphold the reality of imperial Britain's authority, and with the British Isles positioned firmly in the middle, the natural providence of Pax Britannia. There was value in maintaining the imperial world vision, particularly in the minds of the youth (and what an overpowering vision it must have provided). But behind this show of strength was a growing doubt.

The map was one of a number of works published for the Navy League, an imperialist pressure group of politicians, retired sailors and enthusiasts, established in 1894 to promote the continuing centrality of the Royal Navy in the minds of the British and dominion governments. Commissioned by them, and updated at significant cost, it was a valuable reminder in meetings and conferences, educational spaces and in national collections of the particular importance of the Royal Navy in the creation and maintaining of the empire.

The naval glue that bound together this array of territories and dominions, while maintaining the balance of power, was by 1901 spread increasingly thin. Britain's fleet was the world's largest, guaranteed by the 1889 Naval Defence Act (which decreed that it was to be at least as large as the next two biggest fleets combined). But in the face of rapidly expanding German and Japanese navies, choices had to be made over modernisation and where the fleet should be positioned. It

couldn't be everywhere at once, and dividing it reduced its strength, hence the dilemma: was the navy to protect British waters or those of its colonies? Were battleships to be apportioned to each of the 'stations' into which the seas were divided? Choices were required over the exact terms of the replacement of aged ships with the new dreadnought class of battleship.

The plans made by the Admiralty and government were scrutinised and debated very publicly by the Navy League in literature, pamphlets and at events. The League argued for decisions about the Navy to be exempted from traditional party politics, yet provided a popular dimension to the debate that was equally political. In such a high-stakes game the wrong choice had potentially disastrous consequences, the political and naval decision-makers mindful of an empire too large to maintain, and wary of their outdated ships and confident enemies. Britain's answer to her decreasing influence in the Pacific was to forge an alliance with her chief threat there, Japan, in 1902. However, no such admission of the navy's limitations exists in this particular map, or in powerful propaganda later to be produced upon the outbreak of war in 1914 (see 'Hark! Hark!', p. 46).

Others were apparently not so convinced of the long-term survival of Britain's empire through existing naval strength. In a seminal lecture to the Royal Geographical Society in 1904, the geographer Halford Mackinder demonstrated the marginal nature of Britain's maritime power against the secure Central Eurasian 'pivot' zone, which he argued was impregnable from the sea. He was clearly stating that sea power was not the only kind of power; Navy Leaguers in the audience would have been less than amused. Whether or not Mackinder's aim was to arrest complacency and refocus British minds on the defence of empire ('Greater' Britain as well as the British Isles), the point enables an interesting perspective on the world image contained in the Navy League map. For attention truly to be focused upon the value of naval power, surely the seas should have been coloured red instead. Perhaps the map had missed a trick.

1901: Jewish East London: maps, statistical maps and immigration

HERE IS A salutary reminder to approach maps with caution: those that claim a statistical basis are not necessarily neutral. This map illustrates the density of the Jewish population in London's East End, but by focusing on a narrow area of the capital and using heavily nuanced colour-coding, it contrives to be alarmist without actually distorting the underlying data.

Prior to the passing of the Aliens Act in 1905 there were no restrictions on immigration into Britain and there existed, mostly in the major ports and industrial cities, small but thriving communities from around the world. The Act, Britain's first permanent control on immigration, was fuelled by fears of mass Jewish immigration from Eastern Europe. This map, which illustrated a book titled *The Jew in London*, contributed to that debate.[23]

The catalyst was a fresh eruption of pogroms and anti-Semitic violence in the Russian Empire from 1881 onwards. Roughly seven thousand Jewish migrants arrived in Britain annually throughout the 1880s and 1890s. By the time this map was created, the Jewish population of London had reached 135,000 – a highly visible community concentrated in the East End.[24]

Some of the poorest and most densely populated parts of London lay in the East End, including Bethnal Green, Mile End and Stepney. Outside the walls (and jurisdiction) of the City of London and close to the London Docks, they became home to successive groups of French Huguenot and Irish migrants, and are now associated with a thriving Bangladeshi community. According to this particular map of 'Jewish East London', at the turn of the last century it was dominated by one ethnic group.

The Jew in London was one of many contemporary books that considered the 'Jewish question'. Published for the philanthropic Toynbee Trust, it takes the form of two essays, one proposing anti-Jewish legislation, the other strongly averse to it. The map, on the other hand, is anything but balanced. Buildings are coloured to reflect Jewish or non-Jewish occupancy: blue for predominantly Jewish, red for predominantly gentile. Red dominates the map, but given the title it would be easy for a superficial observer to mistakenly assign it to Jewish residents, not non-Jewish.

Even if the map is read correctly, it is still alarmist. Readers then and now could well be reminded of another map, first published twelve years earlier by the social reformer Charles Booth. In his map, which coloured streets by levels of poverty, blue stood for 'vicious, semi-criminal'. Cues such as these distract the reader from highly pertinent 'notes on the map' stating that outlying areas have not been included because they are not as densely populated by Jews. The map therefore creates an unreliable impression of the pattern of Jewish occupancy of the East End. In the 1901 census no more than 18.2 per cent of the Borough of Stepney was foreign-born, of whom fewer still were also Jewish.

Comparable flaws beset Charles Booth's survey of London poverty (for example, on his maps no account is taken of the relatively poor live-in servants in wealthy areas) and the maps are closely related. The detail illustrated shows how the same area of East London was

JEWISH EAST LONDON

SCALE

This Map shows by Colour the proportion of the Jewish population to other residents of East London, street by street, in 1899.

EXPLANATION OF COLOURING.

Proportion of Jews indicated.

95% to 100%.
75% and less than 95%.
50% and less than 75%.
25% and less than 50%.
5% and less than 25%.
Less than 5% of Jews.

NOTE.—In all streets coloured blue the Jews form a majority of the inhabitants; in those coloured red, the Gentiles predominate.

represented in the third series of Charles Booth's *Life and Labour of the People in London* (1902). The streets which, on our first map, were coloured to show the greatest density of Jewish inhabitants are mostly coloured here to indicate 'poverty and comfort (mixed)', the middle category of seven used by Booth for this edition, which descend from 'wealthy' (yellow) to 'lowest class' (black).

Statistics for Booth's map were gleaned from interviews, including those conducted by the London School Board (London's free elementary school system).[25] One of the interviewers, Beatrice Webb (née Potter), produced a number of essays on the working poor for her cousin Charles Booth during the 1880s. She wrote on the clothing trade, in which the Jewish community was strongly represented, then on the Jewish community itself. She was assisted by George Arkell, a statistician and social geographer, who coloured maps according to statistics, and who created our map of Jewish East London. Arkell too had conducted numerous interviews for Booth and toured the poorer streets (escorted by policemen) to make notes.

By the time that Arkell's map was published, pressure from trade unions on behalf of their members complaining about 'sweating' (cheaper wages paid in sweat shops to poorer immigrants) had motivated some Jewish people to agitate for legislation. The perceived threat from the Jews, as articulated, was an economic one – an unsustainable drain on national resources – but this fear of immigration represents a sea change in public opinion.

In the nineteenth century Britain's willingness to accept all comers had often been presented as a source of pride. Much to the annoyance of some of the more autocratic Continental powers, this included granting asylum to numerous anarchists and revolutionaries, from Mazzini to Marx. Technically, 'aliens' had no rights whatsoever. They were present as guests of the monarch, not subjects, and as such could be expelled without warning or explanation. In practice this rarely happened, even when the monarch was in favour of deportations, as Queen Victoria had been after the assassination of Tsar Alexander II in 1881. Britain's liberal traditions took precedence. There was no international war on terror: plotting against the Tsar was not a crime in the UK.

The 1905 Act regulated the situation, officially placing wide discretionary powers in the hands of the Home Secretary rather than the monarch. 'Undesirable aliens' were those likely to place a financial burden on the British taxpayer, either on medical grounds or through lack of means, or those who had been sentenced for non-political crimes abroad. Overall, immigration was reduced. However, for the first time the right to asylum for religious and political refugees (as distinct from migrants) was enshrined in law, and immigration officials were instructed to 'give the benefit of the doubt, where doubt exists'.[26]

It would be a narrow interpretation to consider the Aliens Act in isolation as an example of knee-jerk xenophobia. It marks the beginning of the modern immigration debate, but it was also the most public manifestation of a whole raft of restrictive legislation: the creation of MI5 and MI6, the passing of the Official Secrets Act, the D-Notice system of vetting newspaper stories, the creation of a register of aliens and the routine examination of certain categories of mail.[27] The balance of the right to privacy and freedom of movement, weighed against increased state responsibility – and accompanying state intrusion – in a modern liberal democracy, is rooted in Edwardian unease and paranoia.

MILITARY SKETCH MAP
TO SHOW
SIR R. BULLER'S ADVANCE
FROM CHIEVELEY
TO RELIEVE LADYSMITH
FEBY 14TH TO MARCH 1ST 1900.

1902: The relief of Ladysmith and attempts at rehabilitation

BY CONTRAST TO the appalling maps issued to troops during the second South African (Boer) War (see p. 18), maps illustrating the conflict that were produced for a well-informed British public were often excellent. The London publisher Edward Stanford's 1902 twin maps of the land south of Ladysmith, commissioned and drawn by army personnel, were superb colour lithographic productions. One provided a picturesque topographical description of the region, while the other showed the movement of troops engaged in the relief of the besieged town. It is debatable why anybody, let alone a commercial publisher, would wish to produce a monument to what was at the time regarded as one of the most bungled of British military missions. Yet these celebratory maps prove that not all opinion swung against the commanding officer General Sir Redvers Buller. It is more likely, in fact, that the maps were attempting his and the British Army's rehabilitation during the conflict's darker latter stages.

The outbreak of war in October 1899 had very quickly seen the resident British force outnumbered and hemmed into the towns of Kimberley, Mafeking and Ladysmith with the civilian population. The relief of Ladysmith, from the south, swiftly became the objective of General Buller, who had landed in December with over 20,000 reinforcements. Yet three failed attempts at crossing the River Tulega (the battles of Colenso, Spion Kop and Vaal Krantz) were punctuated by captured guns, confused orders and casualties. Buller was nicknamed the 'Ferryman' in the British press, and when the 118-day siege was eventually lifted in February 1900, the damage had been done.

Relieved of his command, Buller returned to Britain in November 1900 where a struggle for his reputation ensued. It transpired that during the offensive he had sent a telegram to the War Office advising surrender. In October 1901 he was goaded by the *Times* journalist Leo Amery into an impassioned defence of his actions, which involved public criticism of the press and accusations of conspiracy. Amery, along with a young reporter named Winston Churchill and dozens of other reporters, had been an eyewitness at Ladysmith. For his outburst, Buller was sacked outright from the Army.

The press coverage of the war was extensive and initially patriotic, generating massive interest among the British public. Buller had enjoyed a glittering career and was at one time considered for the Army's top job. He was a Victorian hero in the age of the cavalry charge, but he lacked the patience for journalists that was displayed by his colleagues Robert Baden-Powell and Commander-in-Chief Lord Roberts, his great rival.[28]

Buller suffered at the hands of the press, but was exalted by his men. This map was drawn by H. Delmé-Radcliffe of the Welsh Fusiliers, also present at Ladysmith, having been commissioned by his commanding officer, Buller's friend Sir George Barton. Emphasising mitigating factors for Buller's failures (the difficult terrain is delineated in four different ways) may not have been accidental. Set against intrigue and back-stabbing at the top, such a gesture of soldiery solidarity with Buller would not have gone unnoticed. It may not only have been for Buller's rehabilitation that the map appeared in 1902. By then the Boers had been defeated by tactics including concentration camp internment, civilian deaths and 'scorched-earth' burning of farms, all communicated by a now largely horrified press. Diverting attention during a period of crisis was becoming a further weapon of modern warfare.

1905: The Railway Clearing House: South Wales coal and the logistics of rail

BY THE OPENING of the century the coal and rail industries stood at peak production. They were interdependent: railways required coal to power their steam locomotives and the coal industry needed railways to transport it. The spirit of enterprise was not conducive to cooperation, however, and the vast proliferation of railway lines that began to emanate from the South Wales coalfields from the second half of the nineteenth century illustrates the pervading principles of consolidation and extension of control.

Prior to the 1923 grouping of Britain's hundreds of railway companies into just four giant ones, a single journey might involve travelling along lines of different ownership. Imagine the complexity of purchasing a ticket for each stage of the journey. 'Through booking' was made possible, and the lives of passengers and businesses made drastically easier, by the establishment of the independent Railway Clearing House (RCH), set up in 1842 to collect and apportion carriage fees among the many railway companies. This complex task required the employment of an army of office clerks to create and interpret an incredible quantity of minutiae: tariffs for types of freight, regulations on weight and dimensions, and conditions of carriage. These were applied to accurate up-to-date maps of the entire network, which showed patterns of ownership and the chargeable distances in miles and chains.

The most striking of these maps are the Railway Junction Diagrams, produced as handbooks to describe the points where a number of lines merged. Originally conceived and drawn in 1878 by John Airey of the mileage sub-department (who managed to retain copyright of the maps until bought out by the RCH), a number of annotated working copies survive. The hundreds of diagrams are essentially maps of town centres with everything but the railways removed, stripped down with little reference to superfluous features such as terrain. Town centres appear as colourful entanglements, each colour representing a different rail company. Independent and joint-owned lines abut or,

preferably for the companies (who liked to own their own stations), stop just short of each other. For freight, the space in between was the work of marshalling yards.

In the South Wales coalfields, the journey taken by coal was a simple one from collieries to the quayside (and thence to London or the Continent). However, control over these intense routes led to increasingly earnest attempts to bypass rivals. By 1905 South Wales accounted for 18 per cent of all coal mined in the UK, with 16 million tonnes exported annually. Cardiff was then the largest coal port in the world, fed from collieries at the heads of steep valleys to the north. The city's new-found wealth was consolidated by its docks, including Roath Dock opened in 1897 and the 11-mile railway servicing it (coloured purple), which the collieries and feeder railways had no option but to use.

The landscape also played a part. The narrow steep-sidedness of these valleys meant that there was rarely space in them for more than one line. Companies such as the Taff Vale Railway (coloured light green) had complete control over colliery traffic in the Taff, Ely and Rhondda valleys north of Cardiff. But the natural landscape dictated things only up to a point. In 1885 colliery owners from the Rhondda, unhappy with the carriage and dock charges of Cardiff and the Taff Vale Railway, constructed their own vast harbour at Barry and a connecting line which completely bypassed Cardiff and Taff Vale to the west. Such an enormous project could only have been achieved with vast financial resources.

Such was the confidence of the industry that the natural barriers were only partially inconvenient. The Great Western Railway (GWR, coloured yellow) would absorb the Taff and all other local railways in 1923, rendering once crucial stretches of local line surplus to requirements. The jobs of the Railway Clearing House clerks were made simpler at a stroke. By offering direct freight access to London, the GWR moved into direct competition with Cardiff docks and sea transportation.

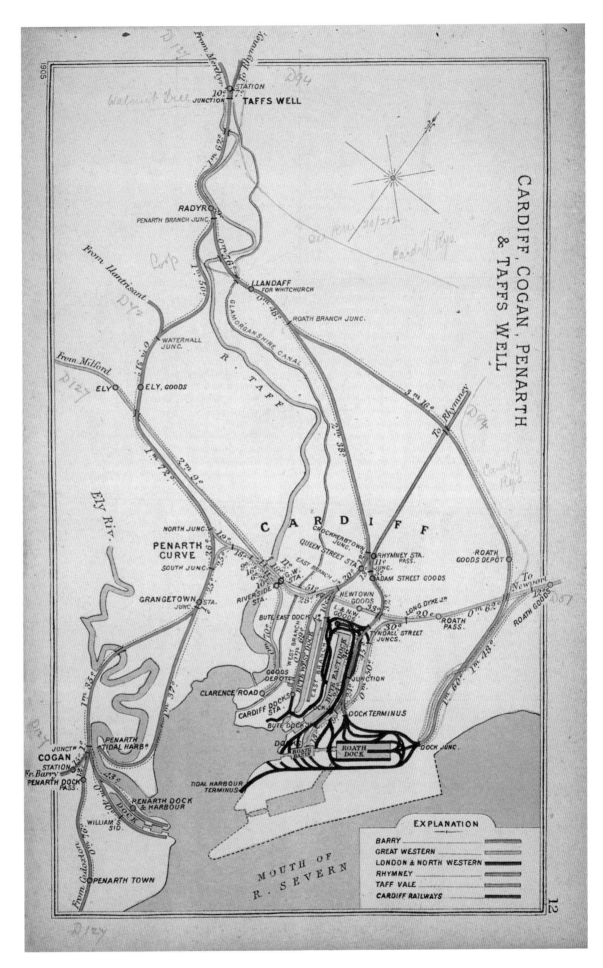

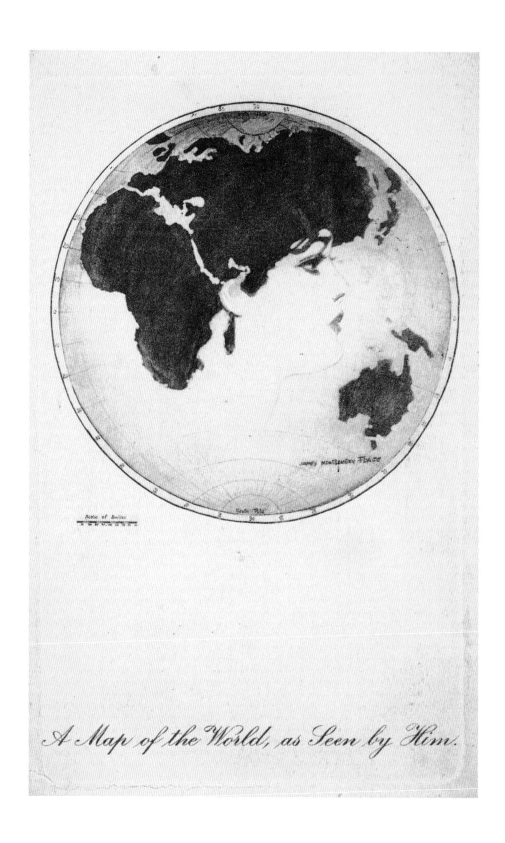

Scale of Smiles

JAMES MONTGOMERY FLAGG

A Map of the World, as Seen by Him.

1907: *A Map of the World, as Seen by Him*: the golden age of the picture postcard

THE EDWARDIAN ERA was the heyday of the picture postcard. Pictorial cards were introduced in 1894, and from 1902 the back of the card was subdivided between address and message, leaving the front entirely free for a striking image. The postcard became a cheap, fast and informal method of communication. Mass literacy and a reliable postal service, which provided up to ten daily deliveries in urban areas, helped to create a social phenomenon which can readily be understood in the age of texting and Twitter. Foreshadowing modern controversies about SMS language, some Edwardian commentators feared that the informality of the medium threatened 'standards'.[29]

Mass literacy extended to maps, as simplified and easily recognisable maps lent themselves to the popular new format, whether as propaganda, holiday greetings or, as here, in an entirely whimsical way. This card was designed by American artist James Montgomery Flagg, soon to become known for his First World War recruiting poster 'I want you for the US Army', in which Uncle Sam takes up the finger-pointing pose made famous by Lord Kitchener. In a gently humorous optical illusion (assisted by a few artistic brush stokes), a map of the eastern hemisphere resolves itself into the face of a beautiful young woman: 'A Map of the World, as Seen by Him'. A besotted suitor sees his beloved's face in everything.

This example was written in 1907, postmarked Woking and sent to an address in Farnham, a town less than 10 miles away. The message reads: 'I wonder if the answer was yes. T. T.'

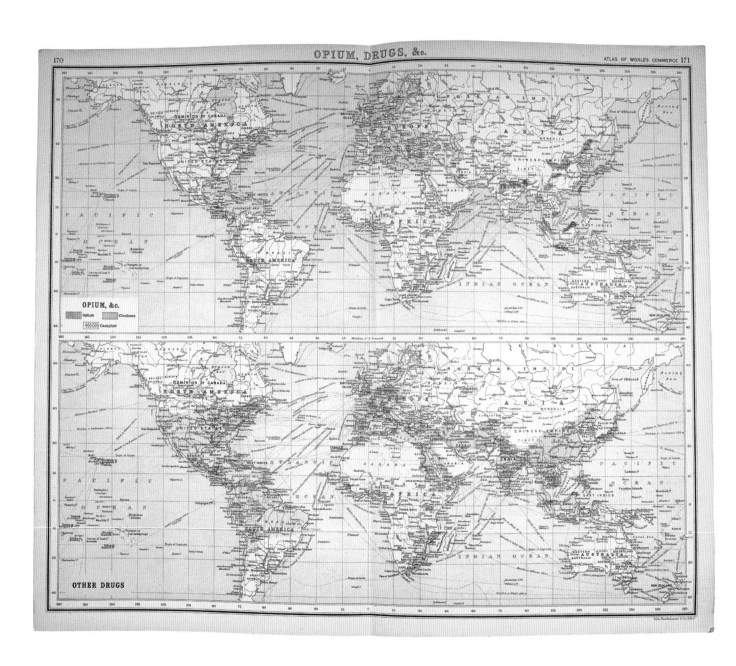

OPIUM, &c.

Opium   Cinchona

Camphor

OTHER DRUGS

1907: 'A pleasant narcotic': opium and the Edwardian drug trade

THIS MAP SHOWING the legal trade in 'Opium, Drugs, &c.' is from Bartholomew's *Atlas of the World's Commerce*.[30] A statistical atlas with a rather dry educational purpose (as well as being a patriotic celebration of Britain's role at the heart of international trade), it offers the modern reader a wonderfully rich snapshot of the Edwardian world. There are maps showing the world trade in every conceivable commodity – from pearls to tobacco, oil, beer, asphalt and wax – and, of course, recording the open trade in commodities such as ivory, furs, ornamental feathers... and opium.

This was towards the end of the era sometimes known as the 'great binge', approximately 1870–1914. Britain was a major player in the opium trade, having won two 'opium wars' which kept the Chinese market open (opium more than balanced out the market in tea and ensured that Chinese silver continued to flow into British hands; British India has been defined as the world's first narco-military state).[31] Even in mid-nineteenth-century Britain there were plenty of objectors to both the wars and the drug, but in this Edwardian atlas opium is still described as 'a pleasant narcotic'. India continued to dominate the market overall, but almost half of British opium was imported from Turkey.

The first Chinese opium dens opened in Britain in the 1860s, shortly after the second opium war.[32] Drugs that are now categorised 'class A' have never been so widely (and legally) available as they were for the next half century: morphine, heroin, opiates of all kinds (and for all ages, including laudanum to quieten children). In mainstream contemporary fiction, the most obvious example of a functioning addict is Sherlock Holmes; an occasional user of morphine, Holmes famously prefers to inject a 7 per cent solution of cocaine, attracting only the occasional remonstrance from the doctor with whom he shares rooms.

What caused the change in public perception? Providing that the 'drug habit' was confined to the upper and middle classes, it could be considered a private weakness. Even drug use by women – an account of 'morphine tea parties' was reported (at second hand) by the *British Medical Journal* in 1902 – was accepted while it seemed not to threaten the social order; the women here were middle class, and they only rolled up their sleeves once the servants had left the room.[33]

Across the Atlantic the situation and perception was very different. Recreational drug use was widespread among large immigrant groups such as Chinese railroad labourers and, increasingly, among the white working class. Cocaine, for example, was popular as a 'work drug' among stevedores. American society believed that drugs threatened to break down barriers of class, race and sex, and the official response was criminalisation. Early US-driven international legislation, such as the 1914 Harrison Narcotic Act, presaged the 1980s 'War on Drugs' conducted by Thatcher and Reagan.

For many Britons in 1914 this remained an exotic, American problem. However, among numerous social controls added to the wartime emergency legislation, the Defence of the Realm Act (DORA), was a 1916 amendment, regulation 40b, which prohibited possession of opium and cocaine in the UK for the first time – a year after restrictions on alcohol consumption were introduced and part of the same drive to keep the nation fighting fit. Morphine and cannabis were excluded, but a precedent was set.

The fatal overdose of actress Billie Carleton after the 1918 Victory Ball at the Royal Albert Hall created one of Britain's first major drugs scandals. Throughout the 1920s the British popular press pushed lurid tales of dope fiends and white slavers: a sinister, highly sexualised underworld to rival America's. Once again, public hysteria and outrage played a greater role in driving drugs underground than medical advice. The two were not necessarily in conflict, but the Great War and its aftermath created a climate of fear: fear of foreigners, that class barriers were collapsing (in both cases expressed through sexual contact), that women were too knowing, too self-sufficient (while simultaneously vulnerable to sexual exploitation). All of these perceived threats to the moral and social fabric of society played a greater role in the initial criminalisation of drugs than concerns about public health.

1908: Philips' model test maps: practical schooling for the heirs of empire

TO AN ELEVEN-YEAR-OLD in 1908, geography lessons would have stood out from the rest of the curriculum. For a start, there would have been a range of colourful and bold images on the walls, such as maps, pictures and new lantern slides. In addition, children had access to illustrated textbooks, globes and 3D-relief or contour models of localities, towns and countries. Pupils were obliged to make their own models out of materials such as sand, clay, Plasticine or soggy paper. From 1898, in fact, model-making became compulsory. 'Learning through play' creativity in the classroom was developed by nineteenth-century Continental educators, who moved away from much criticised teaching methods of rote learning, whereby lists of 'capes and bays' were committed to memory.

The importance of geography as a core subject had risen dramatically in Britain after 1870, the impetus for which was provided partly by the lesson of the Franco-Prussian war of 1870. The Prussian victory was attributed to their better school maps and their greater ability to read them. For the geographical educators, the subject developed independent thought and a geographical imagination, a necessary attribute of the future heirs of empire.

While average eleven-year-olds might not have felt themselves heirs of empire, powerful windows on to the world could now be opened by the huge range of images and literature that had become widely available. These were supplied by a publishing trade wealthy from high demand and cheap production methods. One of the most prolific was the publisher George Philip and Son, who had from the mid-nineteenth century produced a dizzying array of maps, textbooks and models. Perhaps most peculiar of them all were the 'model test maps' of 1905, reprinted in 1908 and afterwards. These maps were black-and-white photographs of real clay models, with the seas coloured and borders drawn around them, without any labels or names. The heights of mountains and widths of rivers were exaggerated, and the coastlines simplified – inaccurate features which drew criticism from geographical writers for providing an 'erroneous impression'.[34]

Such exaggeration was necessary, however, because as their title suggests, these were models upon which to 'test' the accuracy of actual models produced by pupils. Having assembled his or her materials, our eleven-year-old would have crafted a map from memory, or copied the map pinned to the blackboard. If the model was good enough, casts might be made of it as an example to others. Philip & Son also got their money out of their photographed models. The same ones, flaws and all, appear in reduced size next to political maps in *Philips' Model Atlas* (1905) and again in *Philips' Model Geography* (1910), a geographical textbook to be used in conjunction with the maps, containing instructions such as 'draw a sketch map of the river system of India'.[35]

Poor, cheap and inaccurate the 'model test maps' may have been, but to a juvenile their visual power as strong, dark, dramatic images would have been great. Fastened to the blackboard, they presented an exciting (or terrifying) vision of the world. This effect was calculated: geography education was linked to the British imperial world vision and through it everything could be explained. The reasoning had a moral dimension. Britain afforded protection from the wide rivers and impossibly tall mountain ranges of Asia and South America. The world's natural resources were Britain's to extract for benevolent good, and pupils were the heirs to empire with the geographical imagination to enact it.

PHILIPS' SERIES OF MODEL TEST MAPS.

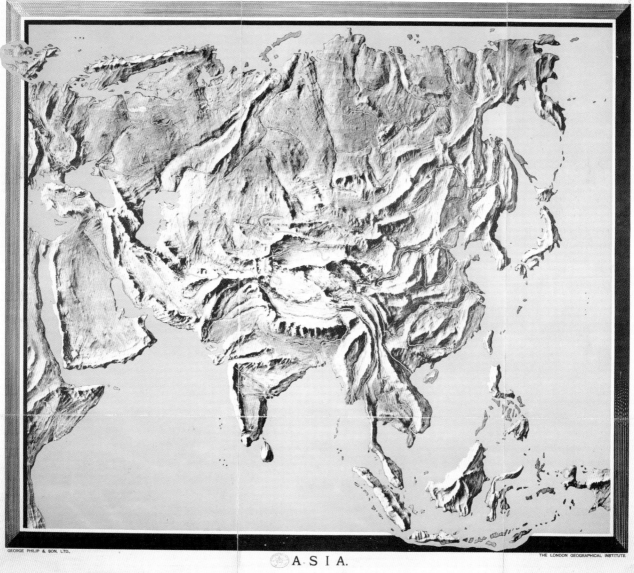

GEORGE PHILIP & SON, LTD.

ASIA.

THE LONDON GEOGRAPHICAL INSTITUTE.

1908: *How to Get There*: the evolution of the London Underground

UR SECOND CARTOGRAPHIC game served as a way of fixing the latest lines of a rapidly expanding network in the public consciousness, and is also an early exercise in London Underground branding. One could always substitute buttons or matchsticks for pennies, but there is a financial aspect to the game and there is no reason to assume that it was intended solely for children. The board is an actual Tube map, officially printed by Waterlow and Sons, which has been pasted on to thin card. A new label – giving the name of the game – has been pasted over the title in the upper margin. Players pick tickets which tell them where they must begin and end their journeys, and the fare they must pay (no more than a penny or two). Players spin a totem to move; landing on (and remembering to call out) 'All change!' entitles a lucky winner to the contents of the booking hall. One might easily come away a shilling or so in profit.

The map itself was printed in 1908, the year that London was hosting the Franco-British Exhibition and its first Olympics next door to one another in purpose-built premises in White City. Three deep-level tube lines opened in 1906–7, so in 1908 the additional coverage was evidently genuinely fresh and exciting enough to warrant turning it into a game. It is also significant in terms of the mapping of the Underground.

London's Underground is the oldest in the world, with stretches of line dating back to 1863. When Harry Beck's famous diagram (an inspiration for so many other diagrammatic subway maps worldwide ever since, from Sydney to New York) was printed for the first time in

1933, he was therefore building on sixty years of mapping experience. Indeed, the January 1933 issue bears the cautious official exhortation 'A new design for an old map. We should welcome your comments'. Beck's achievements are more fully recognised now than they were in his lifetime (his freelance contributions were simply ignored by London Transport after 1960), but it is also useful to appreciate that he was not working from a blank canvas. There are many interesting stages on the road from the earliest maps – often with lines overprinted on detailed street maps of London which were far too large and cumbersome to be used on a station platform – and the convenient pocket-sized tri-fold diagram created by Beck. This is one of them.

The main impetus for change was the creation of the UERL (Underground Electric Railways Company of London), the precursor of the publicly owned London Passenger Transport Board and, ultimately, Transport for London. After a bumpy financial start under the colourful American financier Charles Yerkes, the UERL completed what became the central sections of the modern Northern, Piccadilly and Bakerloo lines. The process of acquisition and consolidation continued through to the formation of the LPTB. It was far from complete in 1908, but for the first time the map gives equal weighting to the four lines which then remained independent of the UERL, and it is regarded as a milestone in branding. The clumsy 'Electric Railways of London' has been replaced with 'UndergrounD', flanked with capitals, which remained a distinctive feature of the logo until the 1960s. The maker of this board game has (forgivably) missed this point completely and pasted over it. However, the map is also significant in other ways. Folding into eight panels, it was much more convenient than its larger predecessors, though not quite as handy as the later tri-fold maps. The surface topography has been thinned out until only the most distinctive features remain, such as parks and main roads, and even they are printed in a subdued monochrome. For the first time London above ground has been relegated to the backdrop for the main show beneath the city streets.

MacDonald Gill and Fred Stingemore would complete the process a decade later, wiping the background blank and even tinkering with the removal of the Thames; that proved a step too far for Londoners, and the river which divides north from south was reinstated before Beck's time. Most important of all, the anonymous, unsung map-maker of the 1908 map hit upon the idea of colour-coding the lines. Many earlier railway maps (such as those issued by the Railway Clearing House, see p. 28) were colour-coded, but until 1908 the Underground had remained resolutely monochrome.

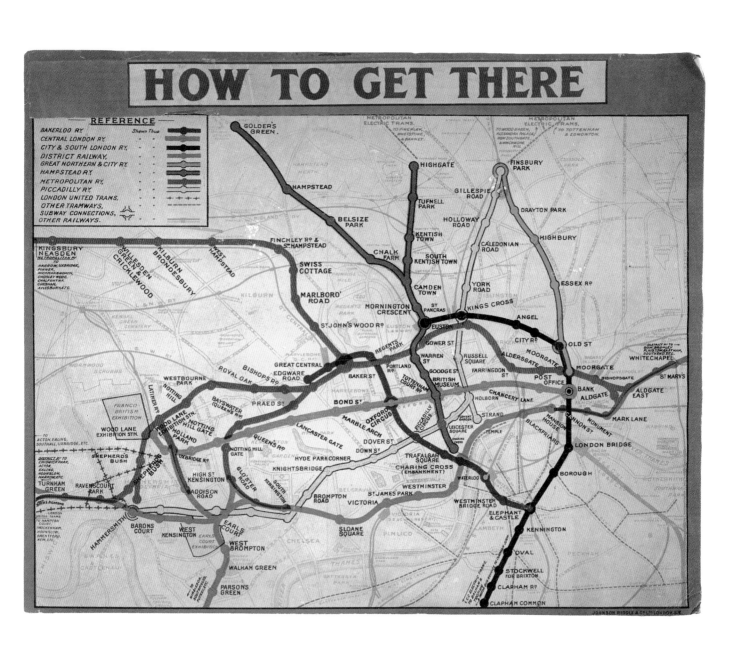

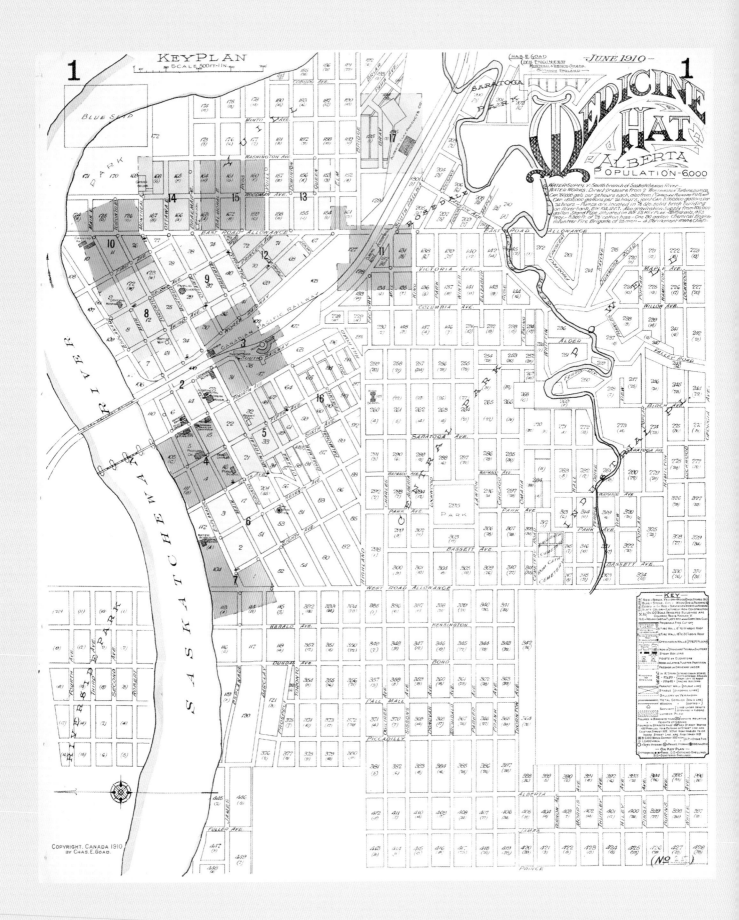

1910: Fire insurance plans: mapping a young Canadian community

MEDICINE HAT IS one of a number of towns in the Canadian Prairie West that was brought into being by the Canadian Pacific Railway (CPR). By 1886 this vast railway, built largely by Chinese and European migrant workers, stretched 1,500 miles from Winnipeg to Vancouver on the Pacific coast. Along it passed minerals, timber, wheat and metals, as well as migrants from Europe, Ukraine, Britain and the United States, encouraged by the Canadian government's promises of prosperity, and settling along the way. In the decade after 1901 the population of the Prairie West tripled from 420,000 to 1,328,000. The Dominion of Canada came into being in 1867, the provinces of Saskatchewan and Alberta joining in 1905. The sense of a nation joined together was passed along the CPR.

The railway made towns, but it also broke them. Some actually uprooted and moved to be closer to it.[36] Along with their development in uniform grids of lots came the institutions and regulations of society. The majority of buildings were constructed using easily available timber, and an inordinate number of them consequently burnt down. It is therefore of little surprise that Canada has an extremely sophisticated fire insurance trade. Some of the earliest companies, in Ottawa and Montreal, were established in the early nineteenth century.

With high regulation came the tools to administer it. The mapping business of Charles Edward Goad, was established in Toronto in 1895. It provided large-scale, up-to-date printed maps of towns for companies to assess insurance risk and to calculate rates. Goad's maps are unparalleled in their detail, and their coverage provides an intimate snapshot of young communities. They mapped each building, its fabric, layouts and fire precautions. They mapped underground storage, and provided details of their occupants. A more authoritarian state could have made much use of this privately sourced information.

Goad's map-makers visited Medicine Hat in June 1910 to update their map of 1905. It must have been a hot day in town, population 6,000 people, heat emanating from the railway sidings, factories and businesses such as Ogilvie's flour mill that had sprung up in order to be serviced by the railway. Photographs show the brand new gleaming facades of Medicine Hat's buildings. Goad's plan shows their interiors. His surveyors went into every building, obtaining information on dimensions, the roofs, windows, skylights, the nature of the businesses and what was stored on the premises. Discussions may have

touched on the skating rink which had coincidentally burnt down that same month. Ten years later, tyre and rubber factories would add extra bustle, heat and fire risk to the town.[37]

Medicine Hat demonstrated to the surveyors a mixture of risk and security. The generous bend of the Saskatchewan River provided close proximity to water, as did the snaking Seven Persons Creek sitting in crude opposition to the ordered layout. Underground water covers were noted, but so were underground tanks of (highly flammable) natural gas that powered the town. The first significant discovery of natural gas had been made in the vicinity of Medicine Hat in the 1890s, and a municipal company was established in 1904 to supply it to homes and businesses.

The semi-arid prairie of southern Alberta suddenly became important. Following the discovery, the CPR employed a geologist to prospect for the oil known to exist alongside it in the ground.[38] One of the founders of the CPR, Donald Smith, Lord Strathcona, was also involved with the Anglo-Persian Oil Company (later British Petroleum), set up in 1909 to supply oil to the Royal Navy which was then searching for an alternative fuel to coal. The bringing of big business to Canada not only improved national unity but imperial unity, as CPR-owned steamers took produce across the Pacific to Asia, and railway developers envisaged vast continent-spanning rail links. Four years later, Canadian troops would cement their unity with empire fighting on the Western Front.

1911: Ulster's prayer: Ireland and Home Rule

AT LEAST AS far back as James Gillray's 1805 etching *The Plumb Pudding in Danger*, in which the world was shown literally carved up on a plate, British satirical illustration had used violence upon the map as a vehicle for political point-making. But if force of portrayal is indicative of the strength of associated passions, then we may view the nonchalance of John Bull's 1911–12 Irish efforts as an unintentionally weak expression. This Belfast-made postcard is Unionist propaganda, produced during the journey of the Liberal government-sponsored Third Home Rule Bill through Parliament. It presents Ireland drifting precariously close to the 'Home Rule Rocks' and away from Britain across the widening 'Union Channel', steadied by a breeched and tail-coated John Bull, his rope passed through Lough Neagh. This tug-of-war could have a number of outcomes.

Irish Home Rule, which meant the restoration of a Dublin government for the first time since its removal by the 1800 Act of Union, had been attempted twice by the Liberals under William Gladstone in 1886 and 1893. Both bills had been defeated by Conservative Unionist (as well as radical Liberal) opposition in Parliament, but ultimately by the House of Lords. 'Ulster will fight and Ulster will be right' had been the battle cry of Randolph Churchill in 1886, attempting to capitalise on the fears of Ulster's Protestants in a largely Catholic Ireland. But in 1911, the ultimate goal of the majority looked close to reality. A Liberal government was again in office, and the 1911 Parliament Act had weakened the powers of the House of Lords to repeatedly block such a bill.

These were the circumstances surrounding the postcard's production, which have been written into the capes and bays of the card: 'Liberal Straits', 'Gulf of Socialism' and 'Molly Maguire Point' (after a pro-Irish American secret society); 'Tariff Reform Channel' references the non-Irish Unionist grievances mixed up in the issue. Protests and rallies were held in Britain and in Ireland. The Conservative leader Andrew Bonar-Law addressed one in Belfast, saying 'I can imagine no length of resistance to which Ulster will go, which I shall not be ready to support'. Intensification of language accentuated the differences between Ulster and the rest of Ireland. The Unionist Ulster Volunteers, a paramilitary group, formed in 1912.

The outcome was neither that Ireland would drift as one on to the 'Home Rule Rocks', nor be grafted on to the British mainland, fitting snugly once more against Lancashire and Wales. More likely, Ulster would be torn away from the rest of Ireland. Ulster, coloured pink just as England and Wales, appears almost inconveniently attached to her three sister provinces. Certainly this was the belief of the Unionist. Ulster was crucially different from the rest of Ireland in that it had industrialised like parts of Britain, setting it apart from the predominantly rural rest of Ireland. The *Titanic* sailed from Belfast's Harland and Wolff shipyard that same year. 'Prosperity Province' was all the Unionists cared about and that is what they got. Home Rule was postponed with the outbreak of war in 1914 and Ulster was severed from Ireland by the Government of Ireland Act (Partition Act) of 1920.

1911: The Delhi Durbar: imperial pomp and pageantry

THIS IS A souvenir handkerchief, printed on ivory silk by the St Martin's Press, London, for the Army & Navy Cooperative Society, a middle-class cooperative that supplied military officers and their families with all their needs, from campaign furniture to groceries. It commemorates the third (and last) Imperial Durbar, the great ceremonial gathering held at Delhi in 1911 to mark the coronation of George V as king of the United Kingdom and emperor of India. He was the only King-Emperor to be present at his own Durbar, wearing his coronation robes and the imperial crown of India, which was created for the occasion. As a piece of imperial theatre, staged to display the military might of the Indian Empire and the allegiance of India's ruling elite, it was a spectacle without parallel.

Maps printed on silk or fine linen have a long pedigree. They were commercially made for tourists – often marketed as 'flexible' pocket maps – or officially produced for use in climates such as India, where heat and humidity meant that cloth could be more durable than paper. Since the eighteenth century they have also been a popular subject for souvenir handkerchiefs.

The likely purchasers doubtless appreciated the attention to detail that has been lavished on the piece. The plan of the Durbar site itself (the elaborate tented city – complete with light railway – which transformed Coronation Park) has been embellished with martial vignettes by military artist Reginald Augustus Wymer (1849–1935) showing soldiers of British and Indian regiments. The varied and distinctive uniforms of the Indian army were a popular subject for prints, cigarette cards and other ephemeral items, and Wymer was noted for

his accuracy. Plan and vignettes are carefully labelled in Hindi, Urdu and English.

In a further display of imperial confidence, the occasion was also used to announce the removal of the capital of the Raj from Calcutta to Delhi. By the time that New Delhi was inaugurated twenty years later, the politcal landscape had changed. However, there is no reason to suppose that the pageantry of 1911 was hubris, intended to compensate for a weakening position.

The Indian contribution to the war effort of 1914–18 would be enormous, in terms of men and matériel. Indian volunteers (including many veterans of the Durbar) fought on all fronts, from Flanders to Mesopotamia. Indian politicians were largely supportive, recruiting and fund-raising. Gandhi, newly returned from South Africa and now actively recruiting frontline troops, declared in 1915: 'I discovered that the British Empire had certain ideals with which I have fallen in love and one of those ideals is that every subject of the British Empire has the freest scope possible for his energy and honour.'[39] The Indian National Congress had been founded on the basis that India needed parliamentary democracy on the Westminster model. Many Indians expected that the sacrifices incurred during four years of war, including inflation and famine at home, would be recognised, and that they would be treated as equal partners in the imperial project. Dominion status, akin to that of Canada or Australia, seemed attainable and desirable.

Indian opinion shifted radically in the immediate post-war period. Through the 1920 map of Amritsar (p. 62) we explore some of the reasons why there would never be a fourth Durbar and why George V's new crown, commissioned at great expense, would never be worn again.

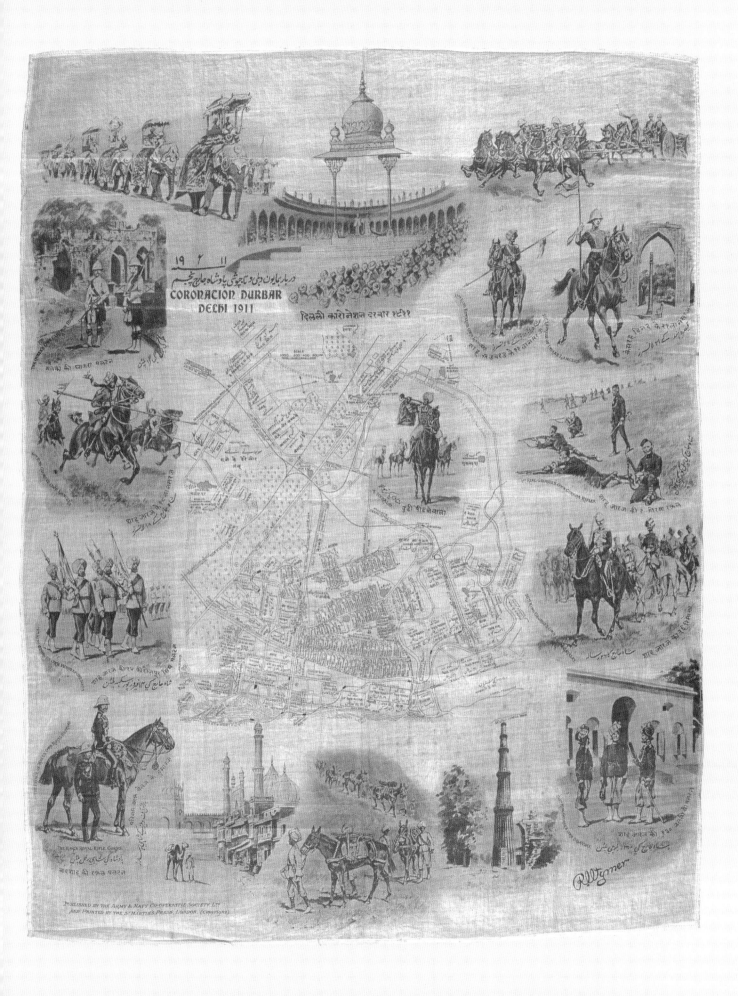

CORONACION DURBAR
DELHI 1911

दिल्ली कोरोनेशन दरबार १९११

1914: Sketch map of an Antarctic adventure: Shackleton's fund-raising appeal

ALTHOUGH BY THE beginning of the twentieth century there were few conquerable parts of the globe remaining, journeys of discovery like those made by Frobisher, Cook and Livingstone remained valuable sources of national prestige and scientific research. All eyes were upon Antarctica. The Norwegian expedition of Roald Amundsen had been the first to reach the South Pole in 1911, just over a month ahead of the British team led by Robert Falcon Scott. Scott and his party perished on the return journey, but for others, the Antarctic held unfinished business.

The explorer Sir Ernest Shackleton had been with Scott on an earlier attempt to reach the Pole in 1901–4. His Nimrod Expedition of 1909 also failed, though achieving the furthest southern latitude of any human. Shackleton was knighted for it – and virtually bankrupted by it – but he was the sort of stubborn, determined and driven man one expects to find at the ends of the earth. As early as 1911 he was planning an expedition to tackle not merely 'the last great Polar journey' but 'the largest and most stirring of all journeys – the crossing of the continent'.[40] Expeditions such as this needed money – and so we come to this map.

At the annual dinner of the London Devonian Association at the Holborn Restaurant on 7 March 1914, Shackleton found himself sitting next to the Member of Parliament for Honiton, Devon, Sir Clive Morrison-Bell. Over the course of the dinner (turbot dieppoise) conversation inevitably turned to Shackleton's Imperial Trans-Antarctic Expedition, planned for that summer. Shackleton was more than happy to explain his plan, and used a blank side of the menu card to sketch a visual aid. He drew as he spoke, outlining Antarctica in a dashed and slightly hesitant series of curves, adding a line with arrows of direction from Patagonia, through the Weddell Sea to Antarctica, drawing a line through the pole to the Ross Sea and on to New Zealand. He answered questions on locating the South Pole with calculations and a further sketch.

The truth was that even two months before embarkation, financial support was far from settled. The Devonian Association dinner and

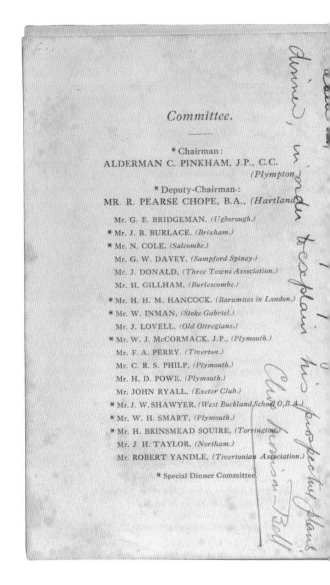

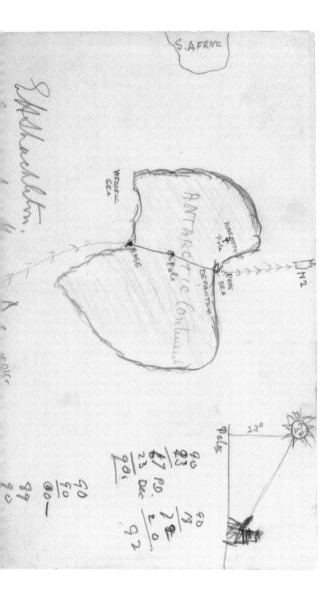

other gala events were perfect fund-raising opportunities. Shackleton, born in County Kildare, was not even a member of the Devonian Association, but Scott, born in Devonport, had been a vice president at the time of his death. The enthusiasm of the group for polar exploration, and any lingering sense of unfinished business among them, may not have been lost on Shackleton, who proposed a toast to the association that evening. Yet he struggled to secure the large funds needed merely through moderately wealthy individuals such as Morrison-Bell. A mass mail-out to several hundred prospects drew only a reasonable response. Finally, in June, his prayers were answered by a cheque for £24,000 from the Dundee businessman Sir James Caird.

The expedition set off on 8 August, days after the outbreak of war, with the approval of Churchill at the Admiralty. The next word from Shackleton came the day before the Battle of Jutland two years later. It is fair to say that the newspapers (at least those that hadn't bought the story rights), as well as the minds of the public, were on the supposedly decisive naval battle rather than Shackleton's latest expensive failure. Shackleton and his party had not crossed the Antarctic. Indeed, they failed even to set foot upon it.

What actually happened was even more extraordinary. The ship *Endurance* became trapped in pack ice in the Weddell Sea far further north than anticipated and was eventually crushed by the ice in November 1915. The party lived on the ice floe until it broke up, eventually reaching a desolate island in April 1916. The entire crew were rescued after Shackleton and three crew struck out for South Georgia Island, successfully travelling 800 miles in an open-top wooden boat.

In the 1930s Morrison-Bell donated the map to the Royal Geographical Society (who were rather frugal in their contribution to Shackleton's cause), having been reminded of it one day when walking past a statue of the great man. Although another cartographic relic of something that never happened, the sketch map is no less insightful of its author, or of the context of polar exploration.

1914: *Hark! Hark! The Dogs Do Bark!*
Aggression and optimism upon the
outbreak of war

UPON THE OUTBREAK of war in August 1914, the language of British political cartoon maps was already developed enough to capture the complexities and culprits of war. This serio-comic map was designed by Johnson, Riddle & Co. in London, probably in September 1914 after the German advance on Paris had been halted by French and British troops first at the battle of Mons (23 August) and then finally at the Marne later in September. From this point, while Russia advanced upon Germany from the east, the opposing armies would 'dig in' along a newly static Western Front. To an enthralled world audience the swift, unpredictable nature of these early stages must have been unbearably exciting and terrifying.

Britain had declared war upon Germany in defence of Belgian neutrality, which had been compromised through the route of the German attack (the Schlieffen Plan). The British bulldog, alongside a French poodle and a wounded Belgian macaw on its shoulder, administers the bites on the nose of the German dachshund, signifying the battles of Mons and Marne. The dachshund, bound inextricably to Austria-Hungary (a mongrel, indicating the empire's ethnic mix), is provoked into action by its numerous troubles: mosquito stings of Balkan unrest (including the assassination of Archduke Ferdinand) and the advancing Russian steamroller. This, in a nutshell, was the official justification for the war and its early conduct. Britain was in fact the junior Ally with only four divisions of troops, compared with France's seventy, but it is easy to appreciate the huge confidence gained from these early successes and how this was emphasised in the press.

'Hark! Hark!' was sold in sheet form and folded into paper wrappers. It presented an artistic, light-hearted and colourful complement to the serious reports contained in the newspaper broadsheets sold in the same shops, which included maps of the fast-moving situation, a couple of days behind events. Purchased and taken home by an educated reader, this map would also have appealed to a child, presenting as it did a pictorial, lucid British account of the hostilities. But whatever dual audience it may have served, the overwhelming impression was that the war would be over soon.

The title refers to the four chief canine protagonists. Dogs are perfect representations of national and human characteristics, enabling the map to sidestep the actual violence being enacted, yet retain something of the savagery. Their inclusion also permitted a contribution from the satirical author Walter Emanuel, whose contributions to *Punch* and books such as *The Dogs of War* (1906), written from a dog's-eye view, added a whimsical light-heartedness to proceedings.

Such was the general optimism of the early war, which was expected to last just six weeks thanks in part to Britain's overwhelming naval superiority, even despite naval setbacks in September 1914. This is emphasised by the giant British sailor–puppeteer holding the strings of ships in the North Sea, Adriatic and Aegean. In the end naval battles would comprise the stalemate at Jutland in 1916. The Dardanelles campaign of 1915 did not even reach the stage of naval assault. By that time the playful optimism of 'Hark! Hark!' had evaporated and Johnson, Riddle & Co. was designing posters for the Parliamentary Recruitment Committee, to attract volunteers into an increasingly humourless conflict.

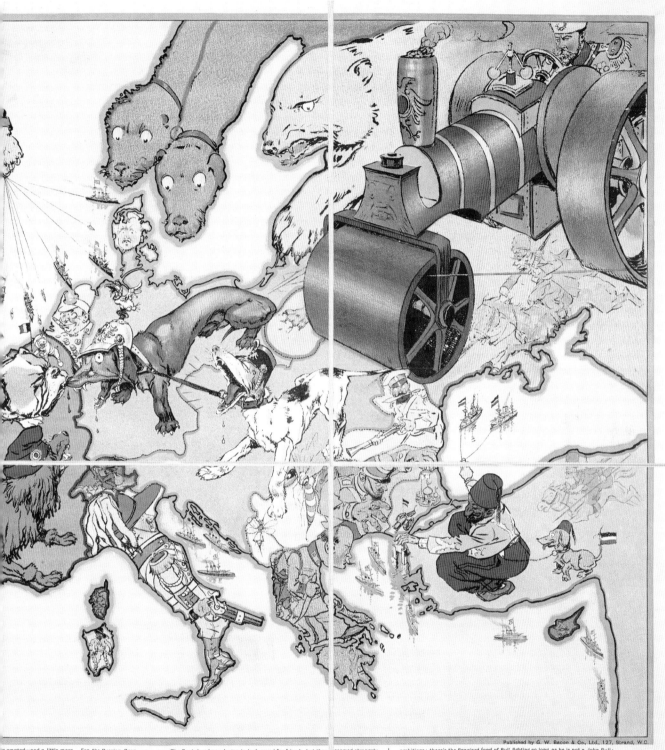

e wanted—and a little more. For the Russian Bear
very game little Belgian Griffon, and there was a
dandified fellow, and there was a Bulldog. Rather
the Dachshund despised him because he was not
t the Bulldog has a habit of sleeping with one eye
grips and won't let go.

attacking the Belgian Griffon, as being the smallest,
elly, but was quite unable to kill her. And he was
nd that the dandified Poodle could fight, and that the
not letting go, and that Russia, after all, was a Rusher.
e Dachshund tremble. And even the little Servian
asty bites, and so did a neighbour of his named Monty.

The Dachshund now began to look round for friends, but they
scarce. He had relied on an Italian Greyhound, a thoroughbred,
Italia dissembled her love in the strangest way, and asserted that
which she could not afford just now. All the same Italia loaded
knows but what it may go off and whom it may hit—for accidents
best regulated families. The Dachshund, to his annoyance, found
and that was a dog of Constantinople. The Dogs of Constantino
known for being fond of offal.

Meanwhile the rest of the European Happy Family looked on,
how the row will spread? There's the Greek with his knife ready
Turkey; there are the Balkans determined not to be baulked o

seemed strangely
named Italia, but
War was a luxury
her gun, and who
will happen in the
only one friend,
ele are quite well

and who shall say
to take a slice of
f their own little

ambitions; there's the Spaniard fond of Bull fighting so long as he is not a John Bull;
there's the Portugee just spoiling for a scrap; there's the Swiss suffering from cold
feet; there's the Dutchman, who keeps smiling with difficulty—still some nice meaty
bones may come his way, and in any event he may be relied upon to play the game and
not to be a Double Dutchman. [ANOTHER NOTE FOR THE IGNORANT :—Holland used to
be known as a low lying country, but this title has now been fliched by Germany.] And,
up North, the Norwegian, the Swede, and the great Dane all have their eyes well skinned.

All this, and more, may be seen depicted above. Search well and you may find
many things. But not Peace. Peace has gone to the Dogs for the present—until a
satisfactory muzzle has been found for that Dachshund. Meanwhile the Dachshund's
heart bleeds for Belgium—and his nose for Great Britain.

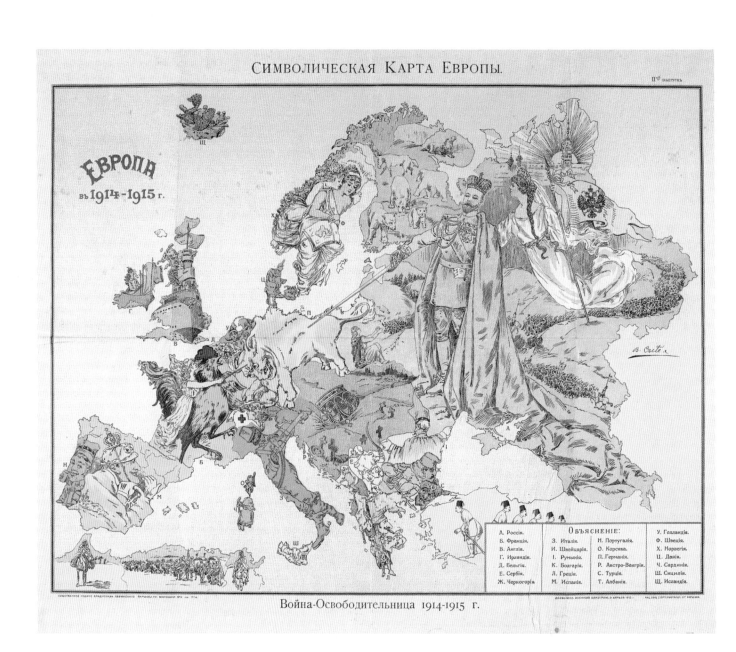

1915: The Great War as seen from Warsaw

THIS 'SYMBOLIC MAP of Europe' by B. Crétée was published in Warsaw by Vladislav Levinsky. It was created as a morale-boosting propaganda piece but, in contrast to the British map 'Hark! Hark!' (p. 46), it offers clues as to how Britain was perceived from abroad during the Great War, in this case by her allies. There is no towering Jack Tar here, although naval might is the key to the image: Britannia is a rather dowdy figure in battleship grey riding a modern dreadnought, while Ireland (depicted as Erin) keeps pace in some sort of fishing smack.

The map was passed by censor on 9 April 1915 and, having been subjected to censorship, it toes the official line. Many caricature maps show Poland struggling to be free, but at the time this map was created Poland had been partitioned for more than a century and Warsaw was the third largest city in the Russian Empire. Poland is personified here as a suppliant, pleading with the Tsar for protection. In maps of this genre Russia is often portrayed as a bear, a steamroller or even a bearded giant, reflecting the size of Russia's army which (it was confidently predicted) would crush the Central Powers by sheer weight of numbers. This map is dominated by the serene figure of the Tsar, pinking a raging German bull without displaying physical exertion of any kind. The Tsar himself is the personification of Russia, and the tsar himself will bring victory.

The origins of the map are neither Russian nor Polish, however. It was originally published in Paris with the title 'Carte Symbolique de l'Europe', firstly under the imprint 'Editions G-D' and the date 1914, and then by Editions Delandre, dated 1914–15, as against expectations the Guerre Liberatrice continued into the new year. Our map is identified in the upper right-hand margin as the second edition and the revisions match those on its French counterpart: Italy entered the war on the Allied side in 1915 and the comic Italian musician has been replaced by a stern soldier; the legend is dropped to the lower margin, creating space to explore the Ottoman role in the war. Crétée was probably a French artist, and this French connection may explain the contrast between a glamorous Marianne riding an especially fine specimen of a cockerel and the grim-faced Britannia.

The artist is fairly kind to the countries on the periphery of the map. Russia's neutral neighbours, Sweden and Norway, are portrayed as two beautiful women in a close embrace. However, in another break from the tradition of anthropomorphic or zoomorphic imagery in cartoon and satirical maps, the Austro-Hungarian Empire is not represented by anything living at all, man or beast. Instead, a fallen crown lies on a barren plain, spotted with graves, a stark prediction of the fall of empires.

This particular genre of comic map-making, showing the countries of Europe at one another's throats at times of crisis, developed in the second half of the nineteenth century. Thomas Onwhyn's Crimean War map of 1854 began a tradition which reached its full flowering in 1914–15, during the extraordinary outpouring of patriotic fervour that greeted the outbreak of war in all the belligerent nations. There are occasional failures of imagination and the national stereotypes that underpin the maps are necessarily crude – they were meant to be interpreted at a glance, without too much explanation. However, they are often also particularly inventive and in this period remained remarkably playful. 'Hark! Hark! The Dogs Do Bark!', for example, presents the Great War as little more than a scrap between dogs. After 1915 the joke seems to have worn thin. There were no cheeky swipes at the horrors of the Somme, Gallipoli or Kut, and there are no examples that can be attributed to the closing phase of the war, referring to the entry of America or the Russian Revolution. When some of these cartographic forms did emerge in later propaganda the results tend to be much angrier, often more visceral, as we will see in the 1942 Vichy propaganda poster (p. 100).

1915: Gallipoli: troop positions at Anzac Cove

THE AUSTRALIAN AND New Zealand Army Corps (Anzac) were one of a number of Allied units to arrive from sea at a point on the Aegean side of the Gallipoli peninsula before dawn on 25 April 1915, and swiftly break through Turkish defences. Subsequently, they were to link up with a combined British and French landing further south and knock out crucial coastal forts guarding the entrance to the Dardanelles straits. This would enable the navy to sail through to Istanbul, which they were to capture, forcing Turkey's surrender and decisively affecting the course of the war. This map, produced by the printing section of the Mediterranean Expeditionary Force (MEF) general headquarters on the island of Imbros, probably in July 1915, illustrates just how far short of the mark these forces fell.[41]

On to the map has been drawn the precise positions of the Anzac bridgehead in red and the Ottoman defenders in green. This is how the situation had remained since April and would continue to do so until evacuation in December 1915 – a stalemate mirroring the stalemate on the Western Front which the Dardanelles campaign had been designed to alleviate.

Prior to the attack, aerial and land reconnaissance by the Royal Navy had picked up the impressively rugged terrain rising up almost straight from the beach of an area immediately north of the designated landing. However, this was inadvertently where the Anzacs landed at around 4:30 a.m. under cover of darkness, as well as enemy fire from the Turkish 2nd Battalion, about a mile further north than planned. The small beach was renamed Anzac Cove, and this name quickly appeared on the new map, together with others created as the offensive developed, such as Walker's Ridge (after Colonel Harold Walker of the New Zealand brigade) and Monash Gully (after the commander of 4th Brigade, John Monash). Further names, such as Dead Man's Field and Bloody Angle, were given to places where the fighting was particularly heavy.

What ensued on the first day was largely defined by terrain so steep that hills such as 'The Sphinx' (Anzacs had been stationed in Cairo) required contours spaced at 200 instead of 40 feet to be properly expressed on this map. The landscape posed incredible difficulties for an attack, especially one robbed of surprise, with the defenders occupying the high ground and able to fall back when necessary. The battle was not swift and decisive as intended, but slow and confused. Caution by Colonel Sinclair-Maclagan commanding 3rd Brigade in the early first wave of attack crucially limited further capacity for advance and enabled time for Turkish reinforcements. When they did arrive, the Turkish 19th Division, led by Lieutenant-Colonel Mustafa Kemal, recaptured the highpoints at the valley heads looking down upon the exposed beach.

These were more or less the same positions in June. They have been drawn on to the map with a hesitant hand, which suggests that the author was copying from another map or a written account. Only one erasure is visible, at the complex series of green lines of Turkish defences that would be named the 'chessboard'. This was the most prominent position, and it was from here that Turkish guns pointed down Monash Valley to the exposed beach, the 'holding pen' where the Anzacs were to stay in the sweltering summer heat (with dense clouds of flies swarming over the ubiquitous tins of apricot jam). Far as they were from effective aid, the conditions for the troops were appalling, with dysentery rife and, as winter approached, frostbite replacing the perils of the heat. The Turkish guns, from which they were never out of range, and the obstinacy of Allied generals hindered evacuation. South of Anzac Cove, the combined British and French landing had fared little better. The very existence of the map, as well as the markings made upon it, would have confirmed to army command the failure of this arm of the offensive.

The tragic futility of continuing the battle, after it was clear that the element of surprise had been lost and that none of the bridgeheads could be expanded even beyond the range of the Turkish artillery, has become part of the identity of Gallipoli to this day. The evacuations of December 1915 and January 1916 were possibly the most successful elements of the campaign. The symbolism has resonated particularly strongly in Australia and New Zealand. The offensive was the first deployment of Anzac troops in the First World War and their contribution relative to the overall size of the formation was enormous. Half of the 410,000-strong British and empire troops involved became casualties. Almost a century later, the names given by them have survived alongside the Turkish ones.

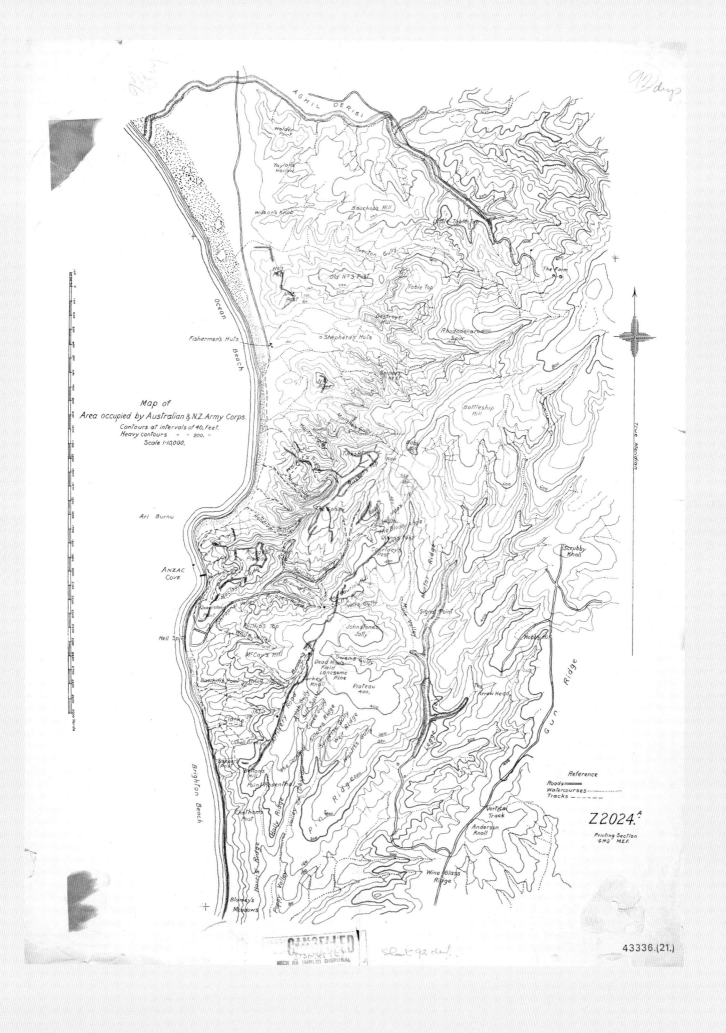

Map of
Area occupied by Australian & N.Z. Army Corps.
Contours at intervals of 40 Feet.
Heavy contours " " 200 "
Scale 1·10,000.

AGHIL DERISI

Ocean Beach

Ari Burnu

ANZAC COVE

Hell Spit

Brighton Beach

Gun Ridge

Reference
Roads
Watercourses
Tracks

Z 2024.ᴬ

Printing Section
G.H.Q M.E.F.

43336.(21.)

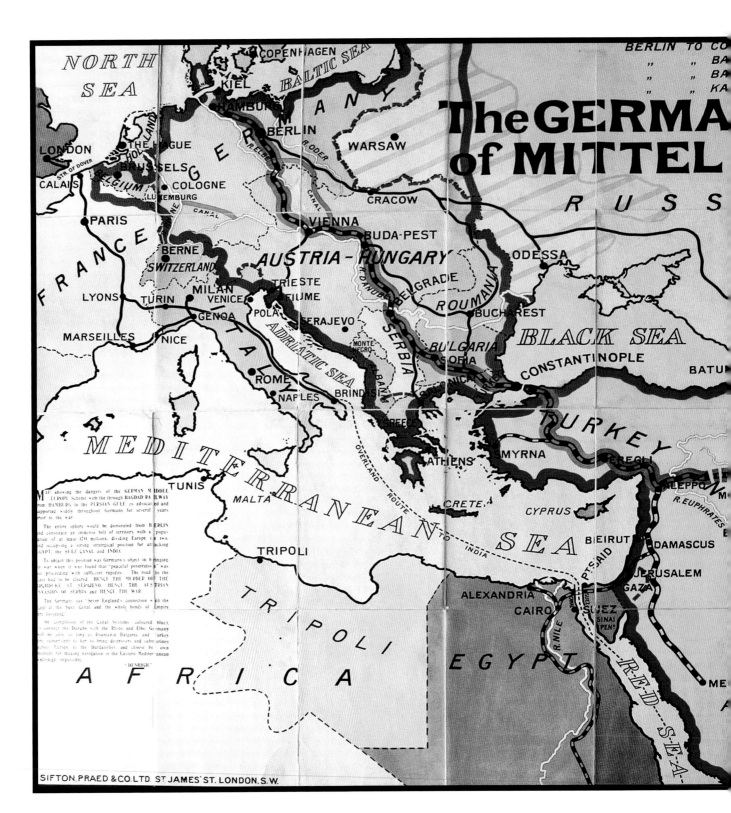

The GERMA of MITTEL

PLE 1150 MILES
2250 „
2950 „
3920 „

SCHEME
ROPA

ASIATIC RUSSIA

CASPIAN SEA

ARAL SEA

BAKU • KRASNAVODSK

TRANS-CASPIAN RAILWAY

• TEHERAN

PERSIA

AFGHANISTAN

BALUCHISTAN

PERSIAN GULF

OMAN

KARACHI

1916: *What Germany Wants*
The Berlin to Baghdad Railway and German ambitions in the Middle East

THE 'GERMAN SCHEME' for 'Mittel Europa' was a theatrical backdrop, measuring an impressive 8 x 5½ feet, commissioned from specialist map-makers Sifton, Praed & Co. to illustrate a morale-boosting lecture by Rudolph Feilding, 9th Earl of Denbigh: 'Why Germany Wanted War, and the Dangers of a German Peace'. The British Empire Union's *Monthly Record* reported on one such delivery of Lord Denbigh's 'striking lecture, illustrated by maps' given to members in July 1918 at the Queen's Hall, a London concert hall, describing it as 'a very successful gathering'.[42]

Its much smaller sibling (overleaf) may well have been given out to the audience. This map does not attempt to reproduce the spectacle of Sifton, Praed's map but conveys similar information in a format suitable for mass personal ownership. It has been 'enlarged and reproduced' from the English translation of the French journalist André Chéradame's *The Pangerman Plot Unmasked*,[43] one of many works warning against German militarism and expansionism. However, Lord Denbigh's 'points to remember' have been transferred from the larger map and reprinted in full on the verso of this one; further copies were available on application to W. H. Smith. It was clearly distributed free and Lord Denbigh's talks would have been one of the obvious venues. If the injunction in the title to 'pin this up in your home' was followed, it would have acted as a permanent reminder of Denbigh's words: why the war was fought and the possible consequences of a German victory or even a negotiated peace.

By 1919 Lord Denbigh had been recruited to the Department of Propaganda in Enemy Countries, located in Crewe House, Mayfair. Aided by journalists and authors including H. G. Wells, he assisted in spreading black propaganda among the soldiers and civilians of the Central Powers.[44] It could be argued that the innovations of the Crewe House Committee were almost too successful, fuelling dangerous myths about the betrayal of frontline German soldiers by elements at home which ultimately contributed to the rise of Nazism and the Second World War. However, these maps are a reminder that they also took domestic propaganda seriously. The text contains a summary of Lord Denbigh's fears regarding German expansion. The panel on the larger map would have taken keen eyes to read from the back of a concert hall, though no one could have missed the German-built Baghdad Railway, a livid scar running from the Baltic to the Indian Ocean. In spite of its tremendous size, the rest of the map has been stripped back to the barest essentials to make the point more starkly.

On the smaller map, too, the bare minimum of detail has been retained; the results are similar, although in this instance the skills of the map-maker have been employed in avoiding all unnecessary

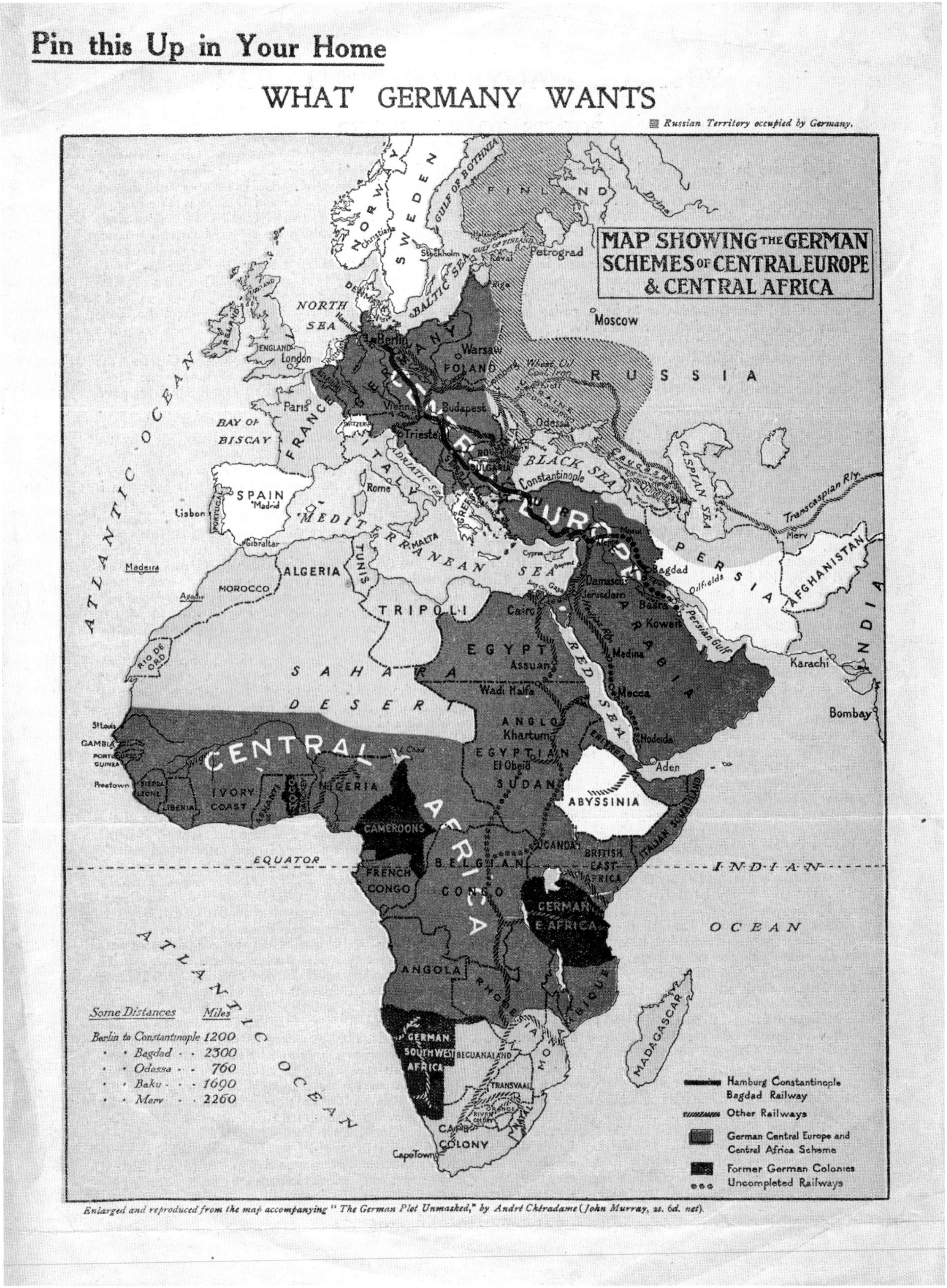
Pin this Up in Your Home
WHAT GERMANY WANTS
Russian Territory occupied by Germany.
MAP SHOWING THE GERMAN SCHEMES OF CENTRAL EUROPE & CENTRAL AFRICA
CENTRAL EUROPE
CENTRAL AFRICA
NORWAY
SWEDEN
FINLAND
GULF OF BOTHNIA
BALTIC SEA
Stockholm
Petrograd
Moscow
NORTH SEA
ENGLAND
London
Berlin
GERMANY
Warsaw
POLAND
RUSSIA
Paris
FRANCE
BAY OF BISCAY
Vienna
Budapest
Trieste
Odessa
ITALY
Rome
BLACK SEA
Constantinople
CASPIAN SEA
Transcaspian Rly.
Merv
SPAIN
Madrid
Lisbon
Gibraltar
MEDITERRANEAN SEA
ALGERIA
TUNIS
MOROCCO
Madeira
Agadir
TRIPOLI
Cyprus
Damascus
Jerusalem
Bagdad
Basra
Kowet
PERSIA
Persian Gulf
AFGHANISTAN
INDIA
Karachi
Bombay
Cairo
EGYPT
Assuan
ARABIA
Medina
Mecca
RED SEA
Hodeida
Aden
SAHARA DESERT
RIO DE ORO
Wadi Halfa
ANGLO EGYPTIAN SUDAN
Khartum
El Obeid
ABYSSINIA
ERITREA
ITALIAN SOMALILAND
St Louis
GAMBIA
PORTUGUESE GUINEA
Freetown
SIERRA LEONE
LIBERIA
IVORY COAST
NIGERIA
L. Chad
CAMEROONS
EQUATOR
FRENCH CONGO
BELGIAN CONGO
UGANDA
BRITISH EAST AFRICA
GERMAN E. AFRICA
INDIAN OCEAN
ATLANTIC OCEAN
ANGOLA
RHODESIA
MOZAMBIQUE
MADAGASCAR
GERMAN SOUTH WEST AFRICA
BECUANALAND
TRANSVAAL
ORANGE RIVER COLONY
NATAL
CAPE COLONY
CapeTown
Some Distances Miles
Berlin to Constantinople 1200
" " Bagdad 2300
" " Odessa 760
" " Baku 1690
" " Merv 2260
Hamburg Constantinople Bagdad Railway
Other Railways
German Central Europe and Central Africa Scheme
Former German Colonies
Uncompleted Railways
Enlarged and reproduced from the map accompanying "The German Plot Unmasked," by André Chéradame (John Murray, 2s. 6d. net).

clutter on a map which could be carried in the hand rather than one which needed to be clear from the back of an auditorium.

The early twentieth century was an age of great unfinished railway schemes: Cape to Cairo and St Petersburg to New York (still a live issue: proposals for a tunnel to bridge the Bering Strait were approved by the Russian government in 2011). The German project – Berlin to Baghdad – was delayed by technical, bureaucratic and diplomatic issues, and even, to an extent, by the brutal displacement of skilled Armenian workers by the Ottoman authorities during the war. Had the line reached Basra ahead of the British and Indian Expeditionary Force that captured the port in late 1914, the course of the First World War could have been very different.[45] If completed, the railway would have consolidated Ottoman control over their diverse and fractured empire, facilitating the movement of men and materiel at speed; it would have granted Germany greater access to the oilfields of the Middle East, and the overland route would have rendered the Royal Navy (which was itself increasingly dependent on that oil) all but impotent in terms of preventing the resupplying of German colonies in Africa and Asia, or neutralising a very real threat to British India. A negotiated peace would have become a real possibility.

Strategic considerations aside, there was another facet to the oriental dreams of the last kaiser. Newly unified Germany came late to the scramble for colonies, but here was a chance to extend German influence informally over a vast region and destabilise the British Empire: a majority of Muslims were subjects of George V. The British and Germans both attempted to harness Zionism and radicalised Islam to their respective war efforts, seemingly with little understanding of any inherent contradictions which that entailed or the possible long-term consequences. The first global jihad – specifically exempting Germans, Austrians and Hungarians and, for the time being, neutral Americans, Dutch and Italians – was a Turco-German initiative. Appreciating the significance of the Sunni–Shia divide within Islam, a German mission to Karbala also persuaded the Grand Mufti to issue a separate declaration of holy war against the English and Russians, with German gold offsetting the threatened revenues from pilgrims from British India. However, the subtleties of the situation eluded the German agents of jihad. They failed to appreciate that German gold could be perceived as a tribute, a form of tax buying a temporary exemption, and that the niceties of such exemptions could easily be swept aside wherever their propaganda took root. They also failed to give any consideration to the poisonous effects it was likely to have in the cosmopolitan but delicately balanced Ottoman world.

The complexities of the situation may have eluded everyone, but the concept of a German-backed pan-Islamic rising would have been familiar to many in Denbigh's audience. John Buchan's popular thriller *Greenmantle*, based on just such a premise, was published in 1916.

Much of the Berlin–Baghdad railway still exists: small Germanic stations dot the countryside in Turkey, Syria and Iraq, though given the fragmented political situation, stretches of the line are seldom used and its original strategic purpose is all but forgotten. However, elements of the ideology which accompanied it have proved more lasting.

1916: A section of the front on the first day of the Somme

O N 1 JULY 1916 a joint British–French attack fell upon German fortified positions along a 16-mile stretch of the Western Front. The front had been static since the end of 1914. Despite meticulous planning over many months, and huge numbers of troops and resources, the first day of the Somme offensive was a disastrous failure. A two-week bombardment followed by mine explosions at zero hour robbed the attack of surprise and failed to destroy the German defences or cut the barbed wire. Instead of punching through the front line, the infantry walked into deadly fire. By the end of the day there were 60,000 British casualties, a third of which were deaths. Half of them came within the first hour of the attack.

As a result, the battle has been one of the most intensely studied and controversial episodes of the twentieth century, with a focus that has transcended the battlefield. Its effects were felt by society. The Somme was the first battle of Kitchener's volunteer army; whole groups of friends who had signed up to fight together died together in the 'Pals' battalions, leaving large gaps in communities and on football terraces. Intense criticism has been levelled at generals for their mistakes, indecision and callousness. Miles from the front line in Army headquarters, removed from the reality of the battlefield, with only maps to recreate the reality of the situation, they might be forgiven for the latter.

More recent interpretations have looked at the conduct of the Great War, of which the Somme was the first full-scale Allied offensive, as a large learning curve in which the lessons and logistics of modern technological warfare were only gradually mastered. One area of this learning curve was improved surveying and map production. Maps informed every level of operations: strategy, the artillery and the infantry. Information gathered by Royal Engineers by means of sound-ranging, flash spotting and aerial reconnaissance from the Western Front was sent to Ordnance Survey in Southampton, who would return printed maps. These would be overprinted with the latest details at corps headquarters or by hand presses in the field.[46]

This particular map was prepared for the 1 July offensive, for the section or platoon commanders in 30th Division battalions on the extreme right of the British attack. Convenient in size, it would have been carried folded along with other heavier equipment which was in many cases discarded. The overprinted red trenches are German; the British trenches were not added in case the maps fell into enemy hands (British troops needed to capture enemy maps in order to find maps of their own trenches). Although a third line of German trenches that had only been spotted by surveillance aircraft in June 1916 was incorporated into maps, the realisation may have come too late to affect the meticulous plan.

It was on this section of the attack that Liverpool and Manchester Pals battalions of 30th Division carved out one of the day's few successes. While others around them fell or held back, they advanced uphill to the German line and captured the fortified village of Montauban. They moved past it and captured the large support trench Montauban Alley. They repelled counter-attacks and, using the map (some of the few on that day actually to have done so), explored the surrounding area, including Mametz Wood, waiting for reinforcements to exploit the gap. And they waited. Three days later they returned to

their own trenches. The cavalry charge, positioned three miles behind the front line, had been cancelled on the morning of the first day.[47]

Back with them may well have come the map. Having reached the possession of Captain John F. Cubbon of the Royal Engineers, it was presented to the Royal United Services Institute in the 1920s. There it sits alongside other salvaged pieces of cartography, which assumed new educational and commemorative roles for the battles that they were produced in their millions to influence.

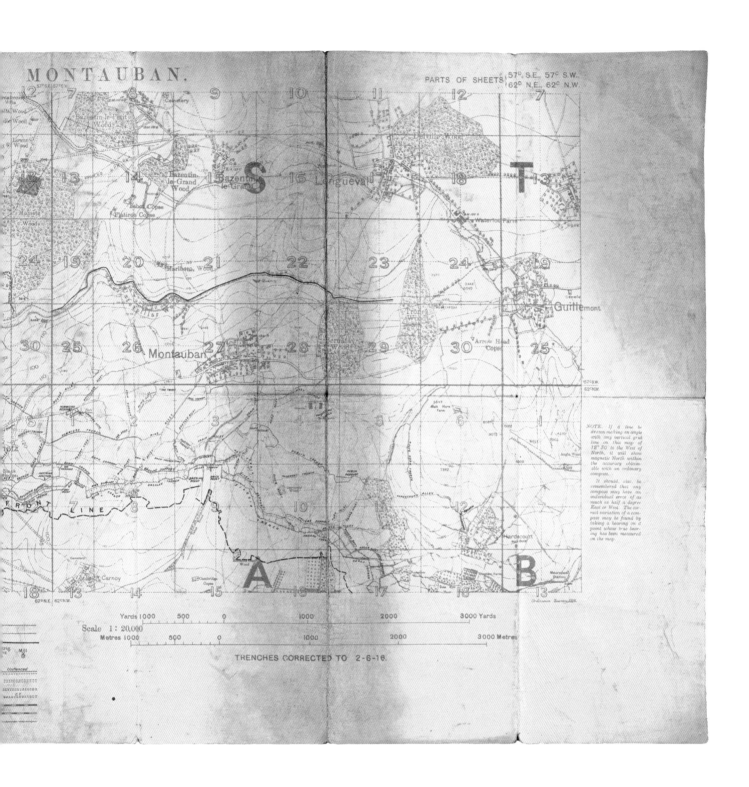

1918: *An Ancient Mappe of Fairyland*: make-believe against the backdrop of Armageddon

THIS *ANCIENT MAPPE* of Fairyland newly discovered and set forth was published the year that the Great War ended, so it is difficult not to relate the two in some way. Could the map constitute a yearning for a return to pre-1914 Edwardian innocence? Compared with the devastated, bomb-blasted landscape of northern France, this vision of a make-believe land may have seemed a seductive escape for a European society bearing the psychological and physical scars of mass conflict. Pinned on to a domestic wall, or folded and shelved in a bookcase (not necessarily just a child's), the map could simply have been a fairy map. However, the war had been so intrusive upon virtually everybody's lives that it is impossible not to view fairyland through this prism.

This map of 'Fairyland', a place loosely defined as one of fantasy, literary imagination and make-believe,[48] is a commanding bird's-eye view of an island with extraordinary topography, populated with a complete range of characters and tales from myth, fable, legend, nursery rhyme, children's story, play and poem. One might, journeying across it, pass through Peter Pan's Never-Never Land to Avalon and the medieval legend of King Arthur, via Puss in Boots, Hansel and Gretel, and Oberon's shield. One of the successful aspects of the map is its convincing space, which enables the eye to pass effortlessly from one story, scene or legend to the next.[49]

If the juxtaposing of Little Boy Blue with Daedalus' Labyrinth seems a little incongruous, it is worth noting that there is a long history of combining different historical literary traditions in maps.

On medieval world maps (*mappae mundi*), for example, various religious, secular, classical, Roman and contemporary worlds are placed together without contradiction. The giants Gog and Magog, and the Trees of the Sun and Moon from the Alexander legend are also often found in these *mappae mundi*. To children, the distinction between the myths, fables and nursery rhymes would have been non-existent.

The bomb-blasted Europe of 1918 experienced a range of complex emotions, including euphoria, anger for some, relief for others, desire for the world to move forward faster, desire for a return to pre-war values, and a coming to terms with the fact that some were never coming back while others had returned physically or mentally disfigured. Had this map been produced, say, in 1900, our perception of it might have been very different. Many of the tales it contains

pre-dated the war and even some of the states that had waged it. However, although the map has continuing and timeless relevance in the fullest sense of fairyland, a proper understanding of the original responses to it is only possible by placing it in the context of 1918.

Bernard Sleigh (1872–1954), a Birmingham-born decorative artist and craftsman, was associated with William Morris and the Pre-Raphaelite Sir Edward Burne-Jones. The map is very much a product of the Arts and Crafts ideology which evinced a return to traditional, pre-industrial production methods. The ornamentation and typeface are in the style of Morris's Kelmscott Press. This retrospective stylistic attitude places the map in opposition to a mechanical modernity, which happened to have reached its most destructive pinnacle during the war.

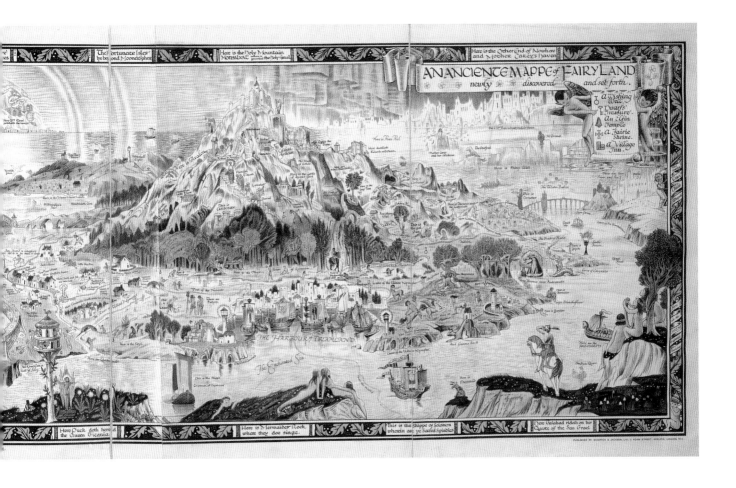

Top Dog! 1919–1945

'We are the top dog!' exclaimed George V with satisfaction on learning of victory in the Great War: a bloodless, narrow summation of the conflict, but to outward appearances not without foundation. The British Empire was about to achieve its greatest territorial extent, through the acquisition of German colonies and former Ottoman territories that were mandated to Britain by the League of Nations as part of the post-war settlement. The USA and fledgling USSR were preoccupied with internal affairs (to all intents and purposes isolationist), and for another generation Britain was able to keep up appearances as the world's only superpower, a perception shared by contemporary figures such as Lenin and Hitler.

However, in 1918 the British world was grieving and sorrow often shaded into anger. Agitation for social and political change had been set aside in the face of a common enemy. Millions had been killed, millions of pounds spent daily to defeat the Central Powers, but in the post-war economic slump Lloyd George's 'fit land for heroes to live in' failed to materialise (see p. 68). In the wider empire, unfulfilled or contradictory wartime promises exacerbated tensions. In Palestine, assurances made to Palestinian Arabs and Jewish settlers under the Sykes–Picot Agreement and the Balfour Declaration proved increasingly at odds, as the 1930 Hope Simpson enquiry discovered (see p. 78). In India, post-war reforms fell far short of nationalist expectations, now higher than ever after the 1917 Montagu Declaration had officially sanctioned the principle of self-government, an acknowledgment of India's contribution to the war effort (see p. 62).

The bloodiest suppression of a demonstration took place in India in 1919, at Amritsar (see p. 62). If there was no second Peterloo to rival Amritsar it may well be an accident of history: the military were routinely called upon to support the civil power in the UK as well as overseas. In the aftermath of the Bolshevik Revolution, and the ensuing civil war that turned parts of Russia into an abattoir, there were calls to bring troops back for home service as rapidly as possible – although there were also concerns about the loyalty of those troops, and fears about what armed gangs of veterans might achieve if they attempted to overthrow the government. Oppressive wartime emergency legislation was extended indefinitely. Echoing the Rowlatt Act (1919), which had contributed so greatly to unrest in India, the UK's own Defence of the Realm Act (1914) was transformed into the Emergency Powers Act (1920). During the Liverpool City Police Strike of 1919 tanks, machine guns and cavalry were used to restore order, while a battleship and two destroyers covered the docks. Mobs were dispersed at bayonet point, shots were fired and at least one man was killed.

Soldiers in the UK were often responding to outbreaks of mob violence, at which warning shots were fired, but few people died. However, as our map showing proposed troop dispositions in London for the 1926 general strike illustrates (p. 68), soldiers were mobilised again

Jerusalem upon your palm.
Jerusalem: Commercial Press, c. 1942

Map cover art is an essential component in the marketing of maps. The care and attention devoted to the outer packaging developed as maps became more cheaply and widely available. There are parallels with the evolution of edition-bound books and the introduction of striking and attractively designed dust jackets. Maps, too, needed to stand out from the crowd. This ingenious cover from British Mandate-era Jerusalem was drawn by F. T. Treitel.

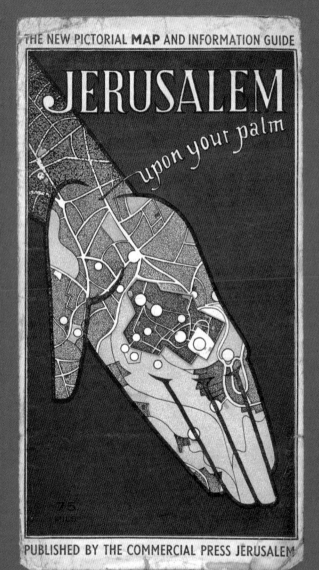

and again to deal with civil unrest in the UK itself; many of those who applauded events in Amritsar would have condoned in principle the use of lethal force on British subjects in the UK. The miners' strike of 1919 was broken not so much on the basis of Lloyd George's vague promise to listen to the miners' demands, as through the threat of bloodshed, and during the miners' strike of 1921 machine guns were mounted on pit heads.

It was an age of glamour and prosperity for a fortunate few, as our map of the new luxury hotel at Gleneagles illustrates (see p. 76), but Britain spent most of the interwar period in recession, which was exacerbated in 1929 by the onset of the Great Depression. It was an age of mass unemployment and political extremism. Britons witnessed the triumph of communism in Soviet Russia and the rise of fascism in Italy, Germany and Spain. The Greco-Turkish War of 1919–22 (p. 64) presented the world with a cruel foretaste of the consequences of redrawing political boundaries along ethnic lines, and it is tempting to see all roads from the 1919 Paris Peace Conference leading inexorably to the Nuremberg Rallies of the 1930s (p. 84), and the outbreak of war in 1939.

The Second World War was Britain's fight for survival. Between the fall of France in June 1940 and America's entry into the war in December 1941, Britain and the empire stood alone against the Axis powers; all of Britain's remaining resources were committed to the struggle, from the bullion in the Bank of England to the iron railings surrounding public parks and squares that were melted down for scrap. In 1945 Britain emerged on the side of the victors, but close to bankruptcy. There was an unquenchable appetite for political and social change, although the privations of the war years segued almost seamlessly into daily life in 'austerity' Britain. Britain's relationship with its empire was on the cusp of change too, as there remained neither the will nor the means to resist strident calls for independence. Continental Europe, shattered by the war, teetered on the brink of economic and social collapse. 1945 also ushered in a period of national myth-making, as most nations (including Britain) sought to rapidly reassess and justify their wartime roles in the light of the new world order. Before the last shots had been fired it was apparent that it would be dominated by the ideological conflict between America and the Soviet Union.

1919: 'A great moral effect': the massacre at Amritsar

THIS DRY, OFFICIAL map was created to explain the events surrounding the Amritsar massacre – a defining moment for the Indian independence movement. It is deceptive in its simplicity: every incident leading up to the massacre is identified, but the Jallianwala Bagh, the public space where hundreds of peaceful demonstrators were killed, is not marked. It was published in 1920 by His Majesty's Stationery Office to illustrate the Hunter Report, the conclusions of the committee of enquiry into the 1919 'disturbances in the Punjab', convened by the Secretary of State for India, Edwin Montagu. Economically overprinted in two colours, symbols denote buildings that had been burnt, destroyed or looted; points where telegraph and telephone wires had been cut; places where people had been murdered and two European women assaulted; and – so crucial in light of what followed – the locations where the commander of the British garrison, General Dyer, had made his proclamation forbidding all public meetings. Every detail leading up to, and suggesting justification for, the events of 13 April 1919 is recorded meticulously. The glaring omission is the Jallianwala Bagh, then a dusty courtyard rather than a garden, a few yards northeast of the Golden Temple.

Punjab's contribution to the war effort had been disproportionately large; the 1917 Montagu Declaration implicitly promised a measure of self-government in return. Under the Government of India Act (1919), Indian officers received the king's commission for the first time, Indians were recruited to the elite Indian Civil Service and Indian women gained the vote, a year after women in Britain. The new system of dyarchy offered Indians greater representation in domestic matters but fell far short of their expectations, and the Rowlatt Act (March 1919), which extended wartime anti-terror legislation and allowed the authorities to imprison suspects without trial for two years, fanned discontent into open unrest. Two of the most vocal opponents in Punjab were spirited away to an unknown location, leading to the escalation of violent disorder recorded on our map.

Amritsar in April 1919 was crowded with visitors from the surrounding countryside attending a fair and celebrating Vaisakhi (an important festival for Sikhs and Hindus alike). The streets were quiet, but there was a sense that they could suddenly slide out of control. Dyer's proclamation banning assemblies was read at various points around the city, but it is uncertain how many people heard it. On learning that a protest was to take place in the Jallianwala Bagh, he set off with a small unit of Gurkha and Baluchi soldiers, fifty of whom were armed with rifles. He also had two armoured cars with mounted machine guns, but the streets were too narrow and he had to abandon them. He had plenty of time to consider what he was about to do, and on arrival in the Bagh Dyer showed no hesitation, ordering his troops to open fire without warning and aim where the crowds were thickest. Approximately 1,650 rounds were fired, most of which hit someone. The official figures reported 387 killed and 1,200 wounded. The actual toll may have been higher: many were crushed to death as they fled through the narrow exits of the Bagh, or took cover in a well. Dyer marched his men back to barracks only when his ammunition was almost exhausted. He made no provision for the wounded: as he told the official enquiry, he did not see it as his job.

In the short term, Dyer achieved his goal: there was no revolution. He claimed to have been seeking to create 'a great moral effect', but the effect was instead to harden the stance of formerly moderate Indians, who now sought complete freedom from British rule. Dyer showed no remorse. The Hunter Commission censured every aspect of his conduct, and stated that 'there was no rebellion which required to be crushed'. British politicians lined up to condemn 'the butcher of Amritsar', from Montagu (who labelled Dyer 'frightful') to Churchill. The response of the British public was more mixed. *The Morning Post* collected £26,000 and commissioned a sword of honour for Dyer,[50] which doubtless cushioned the blow when he was relieved of his duties.

The racial aspects of the massacre cannot be overlooked. Although the incident was unique in scale it was not an isolated atrocity, as Churchill maintained. The authorities routinely deployed soldiers to suppress unrest in the UK as well as overseas. Violent mobs in Britain were dispersed with negligible casualties, but nevertheless the awful human cost of a peaceful protest in Amritsar has skewed modern understanding of this event, which is frequently regarded as purely colonial. Gandhi concluded that empire had corrupted ordinary Britons who exercised imperial power, but it would be more correct to suggest that Amritsar was the most violent expression of a deeper, all-pervasive fear of revolution after 1917.

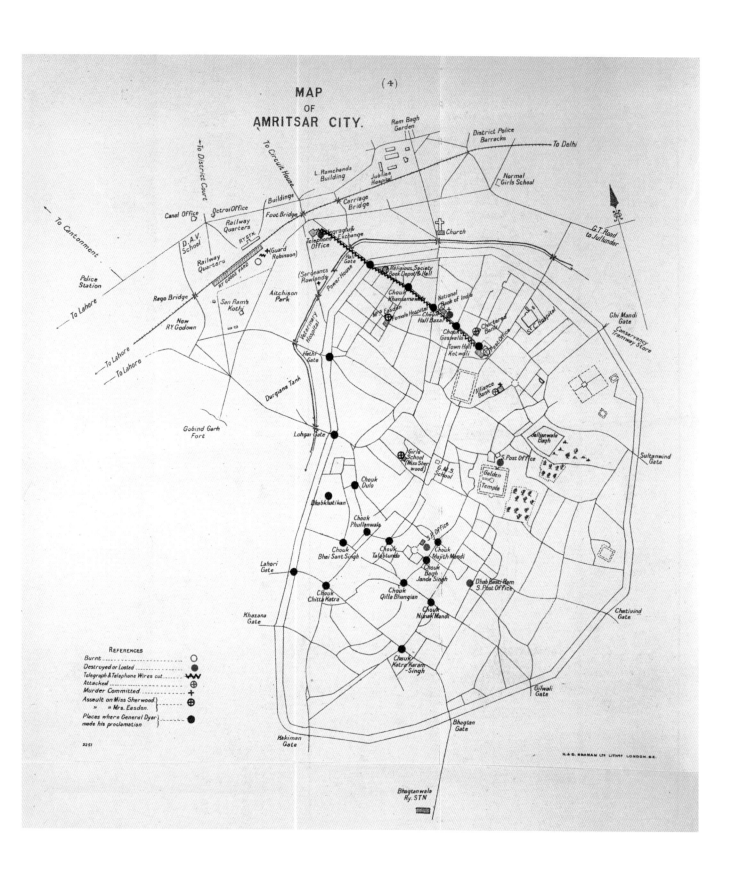

(4)

MAP
OF
AMRITSAR CITY.

REFERENCES

Burnt .... ○
Destroyed or Looted .... ●
Telegraph & Telephone Wires cut ... ✕✕✕
Attacked .... ⊕
Murder Committed .... ✚
Assault on Miss Sherwood ⎫ .... ✚
    „ „ Mrs. Easdon. ⎭
Places where General Dyer ⎫ .... ●
made his proclamation ⎭

3251

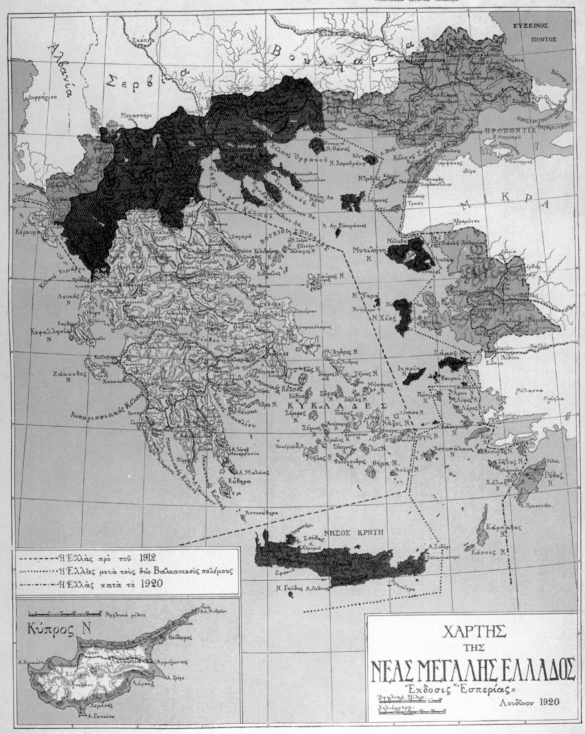

ΧΑΡΤΗΣ ΤΗΣ ΝΕΑΣ ΜΕΓΑΛΗΣ ΕΛΛΑΔΟΣ

Πῶς ὁ Βενιζέλος εὗρε τὴν Ἑλλάδα εἰς τὰ 1910 καὶ πῶς μὲ τὸν ἡρωισμὸν τοῦ Ἑλληνικοῦ στρατοῦ
καὶ τὸν πατριωτισμὸν τοῦ Πανελληνίου τὴν ἔκαμεν εἰς τὰ 1912 - 1913 καὶ εἰς τὰ 1920.

ΤΥΠΟΓΡΑΦΕΙΑ "ΕΣΠΕΡΙΑΣ" ΛΟΝΔΙΝΟΝ.

- - - - Ἡ Ἑλλὰς πρὸ τοῦ 1912
········ Ἡ Ἑλλὰς μετὰ τοὺς δύο Βαλκανικοὺς πολέμους
– - – Ἡ Ἑλλὰς κατὰ τὸ 1920

Κύπρος Ν

ΧΑΡΤΗΣ
ΤΗΣ
ΝΕΑΣ ΜΕΓΑΛΗΣ ΕΛΛΑΔΟΣ
"Ἔκδοσις "Ἑσπερίας"
Λονδῖνον 1920.

1920: Britain, Greater Greece and the Turkification of Asia Minor

THIS 'MAP OF the New Greater Greece' was published in London in 1920, during the Greco-Turkish War of 1919–22. The British were active supporters of the Greeks during the conflict but this map, printed wholly in Greek, was created for the Greek-speaking community in London by the Hesperia Press, which was especially active between 1919 and 1920, publishing political tracts and a weekly Greek-language news-paper *Hesperia* ('western land'). Greece as it was in 1910 is yellow; blue represents the expansion of Greece after the Balkan Wars of 1912–13, but the point of the map is the additional red territory, the 'Greater Greece' of the title.

Greater Greece, the 'Megali' idea, is as old as independent Greece: from the 1830s irredentist Greeks had dreamed of recreating the Byzantine Empire by uniting all territories that could be considered ethnically Greek – which would mean annexing great swathes of Ottoman territory including Constantinople itself. However, this was no pipe dream in 1920. Greek prime minister Eleftherios Venizelos had already presided over one phase of Greek expansion at Ottoman expense in 1912–13, and he now hoped to capitalise on the final break-up of the Ottoman Empire. The demands he made at Versailles in 1919 (see p. 11) were heard.

This map shows the territory that was awarded to Greece at the end of the First World War, as agreed between the victorious Allies and the government of the defeated Ottoman sultan by the Treaty of Sèvres (1920). British troops in Asia Minor handed over their positions to the Greek army. Rather than accept the complete dissolution of the empire and its Turkish heartland, however, army officer Mustafa Kemal (see p. 50) rebelled, founding a rival government in Ankara and defeating Greek and Allied forces in the field. The Allies were forced back to the negotiating table and in 1923 the Treaty of Lausanne recognised the borders of the Republic of Turkey very much as we know them today. The sultanate was abolished months later and the caliphate followed in 1924, by which time the situation depicted on our map had reverted from an internationally recognised reality to the realm of the impossible, though by now the fantasy had become a nightmare. The Megali idea was extinguished by removing the original justification: the Greek community within Asia Minor had all but been eliminated.

The founders of the Turkish Republic were secular, modernising and nationalist. The sultan had ruled myriad peoples, among them Greeks, Armenians, Jews and Assyrians – an extraordinary ethnic diversity born of a rich history which stretched back to antiquity and which was also reflected in the cosmopolitan make-up of the greatest cities of the Ottoman world, Constantinople and Smyrna. Increasingly seen by Turks as the enemy within, all these peoples had been persecuted with great loss of life during the First World War, but the process of 'Turkification' stepped up after 1919. Ethnic and cultural homogeneity was increasingly seen as the cornerstone of a modern Turkish nation state.

For the local Greeks this entailed a murderous scorched-earth policy in the areas shaded red on the map, especially in Thrace and along the shores of the Black Sea and the Aegean. After their army had retreated nearly 750,000 Greeks are estimated to have been killed – approaching half the Greek population of Asia Minor. Some were massacred, but (as had happened already to the Armenians, including those building the Berlin–Baghdad railway: see p. 52) many more died on forced marches – deliberately sent into the barren interior for 'resettlement'. Most of the remaining Greeks fled or were deported, part of the population exchanges with Greece which also saw ethnically Turkish people expelled from their historic homes in what had once been part of the Ottoman Empire.

For Greeks in London, poring over their map of 'Greater Greece', the Megali idea died as Smyrna burned. For us today, this map encapsulates many of the darkest recurring themes of the twentieth century. Even the modest sample of maps in this book bears witness to numerous attempts to partition countries or redraw borders to reflect the ethnicity or religion of a perceived regional majority. Sometimes territorial concessions were not part of the process or were rejected, as here, but the displacement of people on any scale has all too frequently been accompanied by massacres and atrocities as we will see elsewhere in this book. Genocide and ethnic cleansing are two of the ugliest terms coined in the twentieth century, and they have been its constant companions.

1924: A miniature atlas of the British Empire: less is more

QUEEN MARY'S DOLLS' House was a miniature toy model made for the wife of King George V as a gift from the British people. It was exhibited at the British Empire Exhibition in 1924, and is now on display at Windsor Castle. Conceived as a historical snapshot of the British home, and a celebration of British manufacture, culture and elegance, it also offered a light-hearted escapism after the experience of war from which Britain had emerged victorious, but at a price. This incredible object – and the hundreds of accessories made to go in it – does indeed provide a valuable snapshot of how the British monarchy wished to be seen. And although much of it was certainly not 'everyday' for the majority of British subjects, it framed the monarchy in an emphatically new and positive light.

The dolls' house is 1.52 metres high, with over twenty-seven rooms, including a stocked wine cellar, a lift and a strong room for the Crown jewels. It was designed by the architect Sir Edwin Lutyens in the neoclassical style, filled with an array of miniature objects, furniture and artwork that constituted the best of British craftsmanship: a mahogany four-poster bed, a Rolls-Royce car, a bust of Earl Haig, and artwork by John Nash, William Orpen and MacDonald Gill among others. Everyday objects included mops, a Doulton dinner service, a Singer sewing machine and a brass coal scuttle: this latter in the context of imminent miners' strikes (see p. 68) may have appeared almost presciently empty.[51]

The library was filled with original works by some of Britain's greatest authors, a representative collection of literary works that the Dolls' House Committee worked extremely hard to establish. Each major author, each genre and each subject required representation, from Sir Arthur Conan Doyle to Rudyard Kipling. One of the volumes (bound in leather by Sangorski and Sutcliffe) was a bespoke miniature atlas by

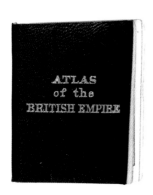

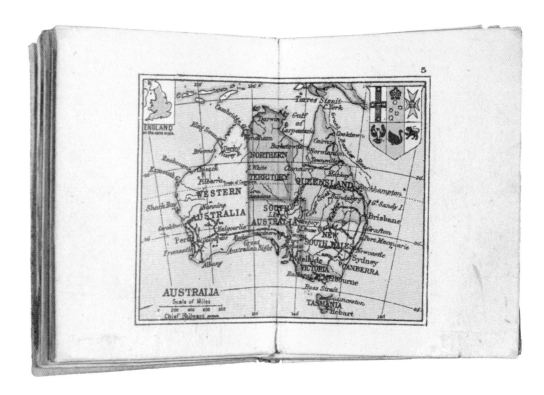

Edward Stanford & Co., containing eight maps of parts of the British Empire. Its inclusion confirmed it as the 'model' geographical work.

Edward Stanford Ltd was one of the most prolific commercial map publishers in Britain at the time. And like the other 'brands', it capitalised on the wonderful publicity gained from showcasing its work in the dolls' house, publishing a replica for general sale. This example was presented to the national map collection, then in the British Museum, but others would no doubt have been purchased by middle-class parents for their children's dolls' houses. This is very significant. Just as the royal family were consciously portrayed through the dolls' house as closer to their subjects by their alignment with everyday objects, those who purchased the atlas were aspiring to greater things. What was good enough for royalty was certainly good enough for them.

The atlas contributed to the well-developed geographical toy market, which included games, puzzles and cards. But it also functioned as a normal atlas in reinforcing the standard, authoritative world view of the British people as prescribed by the establishment. Its contents fix the British colonial interwar gaze: the World, the British Isles, Canada, India, Australia, New Zealand, South Africa, West Africa, East Africa, Anglo-Egyptian Sudan, the West Indies and the Pacific Islands. British possessions were shown coloured red, and British-controlled islands and ports were underlined in red. Moreover, the tiny maps cleverly encapsulated one of the strongest sources of pride underpinning the British Empire through even tinier comparative inset maps of England and Wales to the same scale. The message was emphatically that the geographically small British Isles, by virtue of its industry, ingenuity and moral superiority, had managed to claim inordinately vast swathes of the world for itself. A similar double meaning in terms of scale, and in class, is present throughout Queen Mary's Dolls' House.

1926: A secret map of London: contingency planning for the general strike

ANY SIGNIFICANT INDUSTRIAL dispute today will usually reference the general strike of 3–13 May 1926. This is the national strike by which all others have been measured, huge in scale (involving around 1.5 million workers), with no defined end and, to some, potentially revolutionary in character. The disagreement over pay and working conditions of coal miners led to the Trades Union Congress (TUC) calling out rail and transport workers, as well as those in freight, docks, firefighting and electric power stations. At its root was deep-seated resentment at a lack of social improvement. Trades such as coal mining had been promised better pay and conditions after the sacrifices they had made for the war effort. Returning troops had been promised 'homes fit for heroes' that had not been built. The strike lasted nine days, after which the strikers returned to work, their coal-mining comrades no better off. The Trade Unions Act later in 1926 ensured a strike on such a scale would never be repeated.

This map of London was produced by Ordnance Survey in March 1926 at the behest of the War Office. London might appear far from the mining valleys of South Wales or Durham, and was only one of many cities with sporadic violence (in Liverpool warships kept an eye on proceedings from the mouth of the Mersey). But it was key to the crisis because of the symbolic message an unaffected capital city transmitted to the rest of the UK. The slipping of London into complete anarchy was not an option the government wished to entertain.

They took no chances, buying themselves time to prepare, though disagreeing with the desire of Winston Churchill, Chancellor of the Exchequer, to arm the police. On 31 July 1925 Prime Minister Stanley Baldwin had given the TUC important concessions in the form of a Royal Commission into pay. By the time this commission confirmed the situation of nine months earlier – that miners should take a pay cut – a contingency plan was in place which included an Emergency Committee on Supply and Transport, a framework for attracting volunteers, and the production of aids such as maps. The War Office copy was 'stamped' as an accession the day after the commission had sat.

It indicates the seriousness that the government attached to the threat, and how thoroughly they had prepared for it. The most heavily defended areas – those with the most symbols – are the financial, administrative and commercial centres of the city. But, in addition to the explanation of symbols for rail, roads and boundaries, the key provides an explanation of the extra 'secret information': the whereabouts of barracks, territorial units, fire stations and 'vulnerable points', such as the temporary milk depot to be set up in Hyde Park.

It is important to recognise that over-emphasising the map's confidentiality (it has been labelled 'secret', which was later downgraded to 'confidential') reflects not so much the sensitivity of the information as the fears of the government. This caution was manifested in other government preparations.[52] Any publicity of the existence of this map would have shown the establishment to be fearful and have had a detrimental effect on morale. In short, it wasn't the contents of the map, but its very existence, that was most sensitive. It was not commonly known about until 1995.

1926: The Hundred Acre Wood

E H. SHEPARD'S map of the Hundred Acre Wood, first published on the endpapers of A. A. Milne's *Winnie-the-Pooh* in 1926, has introduced generations of children to the concept of cartography. The first edition was printed in 35,000 copies, of which 3,000 were bound in red, blue or green leather for bibliophiles. The first collection of Pooh stories was an instant success with adults and children, and it has never been out of print since.

The setting was inspired by Ashdown Forest in Sussex, near Milne's home. Milne took his illustrator to explore 'all the spots where the things happened'.[53] Shepard's map, ostensibly drawn by Christopher Robin ('and Mr Shepard helpd') stands within the genre of literary and fantasy maps discussed elsewhere, such as Sleigh's 'Mappe of Fairyland' (p. 58) and Tolkien's maps of Middle Earth (p. 132). However, this is specifically an adult interpretation of a child's universe. It is an effective guide to the events of the book. The juvenile reader can refer to it to see where the houses of Pooh, Piglet, Kanga, Rabbit and Owl stand in relation to one another, not to mention 'Eeyore's Gloomy Place, rather Boggy and Sad'. Other key spots are also identified, such as the location of the 'Pooh Trap for Heffalumps' and the 'Bee Tree'. The faux naïf orthography and artfully childish misspellings are supported by the subversion of cartographic conventions: the compass rose, for example, spells out 'POOH', as Christopher Robin has no need to know his north from his south.

Milne and Shepard had both contributed to *Punch*, and in 1924 they had collaborated on Milne's book of poems for children: *When We Were Very Young*. Milne recognised that Shepard's illustrations, including his map, were an integral part of Winnie-the-Pooh's success, and in an unusual move he arranged for Shepard to have a share of the royalties. Shepard, for his part, ultimately resented his close association with 'that silly old bear':[54] like Milne's son, the real Christopher Robin, he came to feel that Pooh had overshadowed his life.

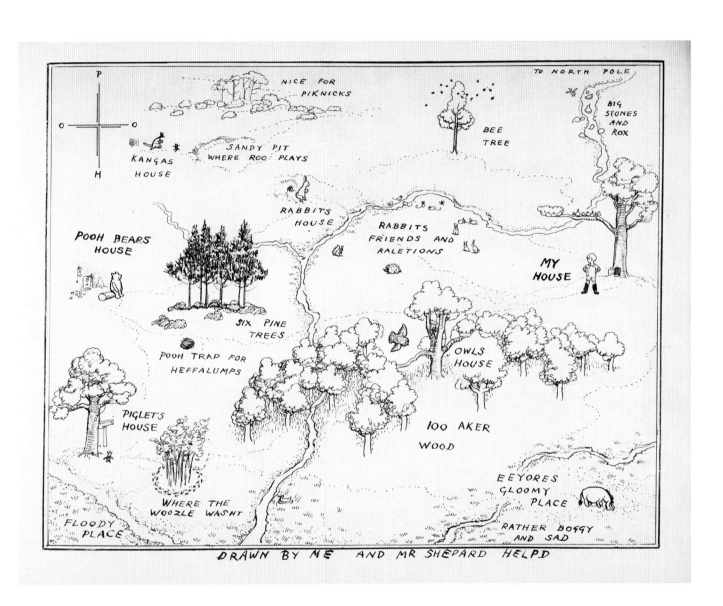

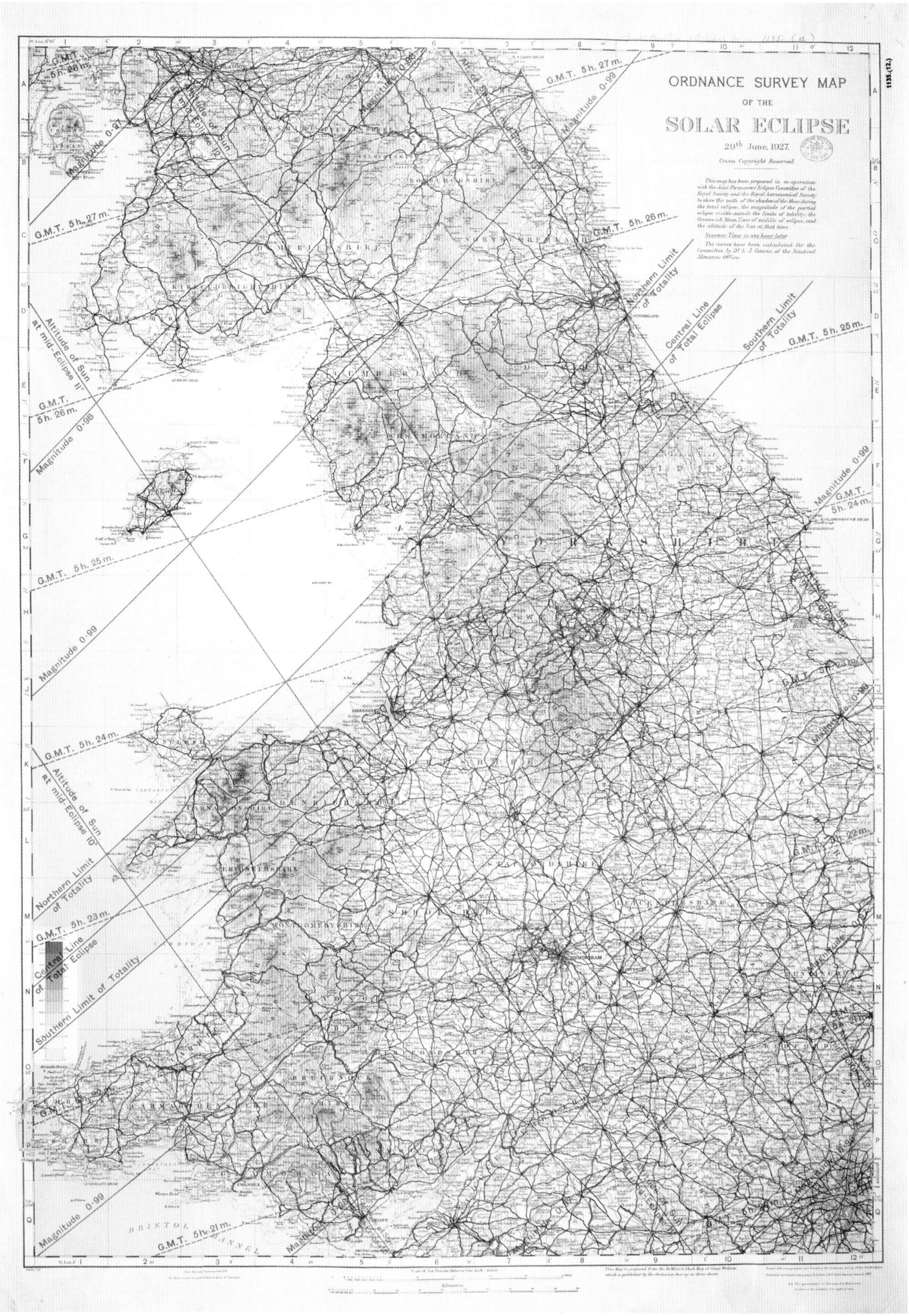

ORDNANCE SURVEY MAP
OF THE
SOLAR ECLIPSE
29th June, 1927.
Crown Copyright Reserved
This map has been prepared in co-operation with the Joint Permanent Eclipse Committee of the Royal Society and the Royal Astronomical Society to show the path of the shadow of the Moon during the total eclipse, the magnitude of the partial eclipse visible outside the limits of totality, the Greenwich Mean Time of middle of eclipse, and the altitude of the Sun at that time.
Summer Time is one hour later
The curves have been calculated for the Committee by Dr. L. J. Comrie, of the Nautical Almanac Office.
Northern Limit of Totality
Central Line of Total Eclipse
Southern Limit of Totality
G.M.T. 5h. 27m.
G.M.T. 5h. 26m.
G.M.T. 5h. 25m.
G.M.T. 5h. 24m.
G.M.T. 5h. 23m.
G.M.T. 5h. 21m.
Magnitude 0·99
Magnitude 0·98
Magnitude 0·97
Altitude of Sun at mid-Eclipse 11°
Altitude of Sun at mid-Eclipse 10°
BRISTOL CHANNEL

1927: Celestial bodies and holiday traffic: a cartographic guide to the solar eclipse

THE SOLAR ECLIPSE of June 1927 caught the public imagination. It was the first time in 203 years that a total eclipse of the sun was visible to parts of the UK. Media and commercial interest – and the opportunity for a holiday – produced an atmosphere akin to national celebration. The majority stayed at home, since the eclipse would have been at least partially visible to all in the UK. However, those lucky enough to inhabit the zone of totality would have witnessed the moon completely obscuring the sun. In the days leading up to zero hour on 29 June, thousands of people travelled to the strip of Britain stretching from rural north Wales to Lancashire and the Yorkshire Dales. An estimated 200,000 people travelled on trains, including 'eclipse special' services chartered to various places including 'eclipse town' itself, Southport. Many others travelled by car, and late June 1927 saw the first example of modern holiday traffic congestion.

One of the main ways that people understood the solar eclipse was through maps. Diagrams in public information leaflets and newspapers illustrated the line-of-sight concept. Maps of the UK with a dark diagonal line through north Wales, Lancashire and Yorkshire illustrated the zone of totality. Similarly crude maps were extremely popular in advertising. Town councils, transport and travel companies, hoteliers and manufacturers all smelt the money in such public interest, and adverts for products with maps bearing slogans such as 'not to be shaded by the eclipse' and 'A total eclipse of all others' proliferated.[55] Seizing the opportunity with its new commercial thrust, the Ordnance Survey produced a special souvenir map in early 1926 with the Royal Society and Royal Astronomical Society. A practical guide and a memento, the 'eclipse' map was a fine production and, at 3 shillings, correspondingly expensive. It showed most of Britain around the shaded eclipse zone, at a scale of 10 miles to the inch. The cover bore a dramatic eclipse scene by Ellis Martin.[56] The overlaid eclipse data was impressive: a slightly shaded area for the zone, the track of the eclipse with times, and the magnitude and altitude of the sun.

For those driving, the map would have been a useful way-finding tool, though only up to a point. While it showed how to get to the destination of total eclipse, on the evening and night of 28 June, when many travelled, it may have been of less use in finding temporary paid viewing sites not marked.

It may not even have guided some to totality, for in an unfortunate publicity incident for the Ordnance Survey, official observations of the sun in May had altered the track of the eclipse north-west by 1 mile. It also put the OS data four seconds out. Letters and complaints in the local and national press drew the response that 'on any map showing the zone of totality it would be unwise to trust to the edge of the shadow being right within half a mile'.[57] Following the map to the letter, it is easy to feel sorry for those who stationed themselves a few feet within the zone, or looked away a second too early. Although the OS map was far better in terms of accuracy than the poor advertisement and newspaper maps, by its own exacting standards it had been found wanting.

Previously, celestial mechanics had been the preserve of professional scientists and astronomers. But with the advent of mass media, cinema (film crews covered the eclipse), the BBC (10 million Britons owned a wireless) and special public information booklets, everyone was capable of believing themselves proficient in the sciences. In true British fashion, the cloudy weather meant that many did not witness the total eclipse at all. But also in true British fashion, according to recollections and reports in the press, amid the cloud, congestion and rain, everyone gritted their teeth and enjoyed it.

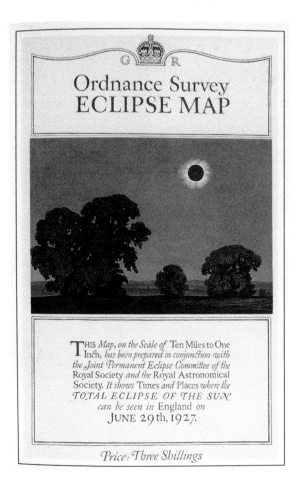

1929: A pictorial chart of English literature for the American tourist

THIS PICTORIAL CHART of English literature was published by the Chicago map-maker Rand McNally for American literary enthusiasts and schoolchildren. England, Wales, and snippets of Ireland and Scotland are shown festooned with the names and portraits of their most famous authors, and the occasional famous scene brought to life. William Shakespeare, Charles Dickens and Robert Burns are placed where they are purported to have lived and worked. For American tourists to Europe, who had for over a century been rediscovering their British heritage and shared culture of common language, this was an itinerary map. But it also acted as a surrogate pilgrimage for those who wouldn't make the journey across the Atlantic. As such, it presented an altered version of reality, a theme park Britain of the Disneyland mould (see Walt Disney World, p. 206) that fed and nurtured a particularly American view of Britain. This, of course, was entirely consistent with the blend of fiction and reality contained within the pages of literary masterpieces.

That is not to say that the 'hordes of Americans seeking the romantic experience of literary encounter' did not have their own home-grown heroes – they were introduced to Britain by them. Washington Irving, for example, had travelled in Britain in the 1820s and described his experiences compellingly. The novels of the Brontë sisters and Jane Austen, which were widely read from childhood, also shaped Americans' perception of Britain. As the heirs to British culture, such affinities between British and Americans were there to be reclaimed.[58] Throughout the 1920s we can discern the overwhelming confidence of the United States and its tourist industry, when the dollar was strong and European excursions were common and easy.

Yet this snapshot of the twentieth-century literary landscape does not correspond with our modern impression of the 1920s, which was such an exciting and subversive period in literature. This is a pre-modern topography, the avant-garde nowhere to be seen. The name of James Joyce, for example, Dublin's now-favourite son, is noticeably absent alongside Yeats, Shaw and Swift. Joyce was at this point in Paris with America's lost generation – Ernest Hemingway, Henry Miller and Gertrude Stein – today regarded as America's greatest twentieth-century writers. There is no bearded D. H. Lawrence hidden amid the Nottinghamshire wool mills. Certainly, it is difficult to envisage his obscene 1928 novel *Lady Chatterley's Lover* in a school, let alone in the English giftshop alongside the tea and scones of tourist-attracting Britain.

But while the disreputable modern landscape is banished from the literary map, other older ones are admitted: it depicts not simply romantic or Victorian literary Britain, but Anglo-Saxon, medieval and Elizabethan. The Spanish Armada (1588) sails on a trajectory with a Viking longboat (the first Viking raid was in 793). Around the edge of the map are images of British history that could be straight out of the popular 1930 historical satire *1066 and All That* (Ella Wall Van Leer's illustrations contain more than an aesthetic similarity with those of the book's illustrator John Reynolds). Just as that celebrated school history spoof intentionally blended fact with fiction, the same recipe of real and imaginary is evident in Rand McNally's chart.

This is literature in its truest form, for the tourist–pilgrim–student user who did not discern fact from fiction. The difference between the two is imperceptible. J. R. R. Tolkien was to claim the same when justifying the authenticity of his created English mythology (see his map of Middle Earth, p. 132). Thomas Hardy created a fictional land called Wessex which was so ostensibly real (and commercially lucrative abroad) that he became its sole historian.[59] When, in 1990, the Ordnance Survey of Great Britain added 'Doone Valley' into their official mapping of Exmoor in Devon, in homage to R. D. Blackmore's 1869 novel *Lorna Doone*, the American Lorna Doone Society objected to its inaccurate positioning, showing just how arbitrary the division between fiction and reality can be under the scrutiny of experts.[60]

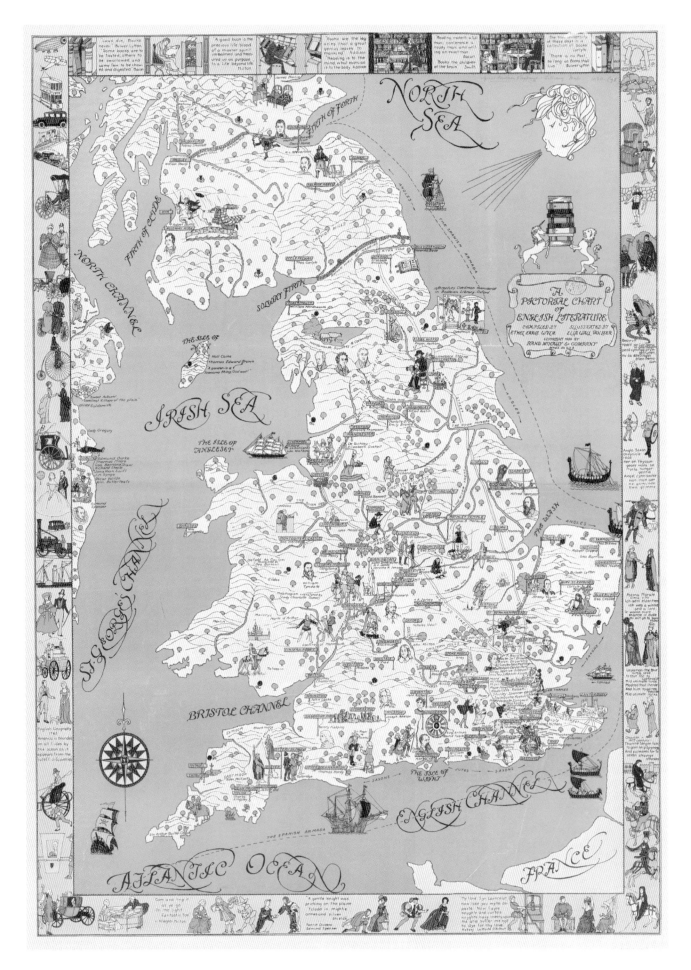

1930: The Gleneagles map of the heart of Scotland

THE GLENEAGLES HOTEL was one of the last great railway hotels. Opened in 1924, it was operated by the London, Midland and Scottish Railway Company, one of the 'big four' British railway companies that emerged from the grouping of the previous year. Railway heraldry is emblazoned on the lower corners of the map, beyond the oviform border.

Unlike its nineteenth-century predecessors, Gleneagles was not located next to a major terminus, though it was served by a small local station. The hotel was a destination in itself, known for its luxurious surroundings and its golf course. In 1930 *Vanity Fair* enthused: 'Somebody once called Gleneagles Hotel his "Palace in Scotland". And what a palace it is! A palace set among the heather and the furze and the hills, where society and business people from both sides of the Atlantic come to take their share of the sport and hospitality of this famous hotel.'

Should the guests tire of golf or shooting, a copy of this map could be purchased for half a crown, mounted on linen and folding into protective covers. This is a map for travelling, showing places of interest within easy reach of the hotel by road or rail. Some of the references are literary: 'Rob Roy Country' and 'Lady of the Lake Country' are drawn from the works of Sir Walter Scott, and a dotted line marks the 'probable' route taken by David Balfour in Robert Louis Stevenson's *Kidnapped* (1886). Others are historical: the site of the Battle of Bannockburn, Glamis Castle and Queen Victoria's favourite view over Loch Tummel. There are engineering marvels, such as the Forth Bridge, other hotels and hydros, and sites of natural beauty. This is a romantic view of the heart of Scotland for the wealthy tourist, with Gleneagles at its core.

1930: Immigration and land settlement in Mandatory Palestine

THIS IS AN officially produced map showing land purchases by Jewish organisations in British Mandate-era Palestine. The base map is the GSGS (War Office) survey compiled by the Survey of Palestine, and it was appended to the Hope Simpson *Report on Immigration, Land Settlement and Development in Palestine*, published by HMSO in 1930. A Royal Commission, chaired by British politician Sir John Hope Simpson, was established in the aftermath of the Palestine riots of 1929, and Jewish immigration was identified as one of the root causes.

The British had captured Palestine from Ottoman Turkey during the First World War (see p. 60), and (in common with a number of other territories that had been controlled by the defeated Central Powers) it was mandated to Britain by the League of Nations, to be administered until such time as it could become self-governing. British rule was legitimised on a temporary but open-ended basis. The Ottoman Empire had been generally hostile to Zionist projects to boost Jewish settlement in the region, and the Jewish population had fallen during the war. The British, on the other hand, had made a public commitment to the creation of a Jewish homeland through the 1917 Balfour Declaration. However, the strength of local Arab feeling against any increased Jewish presence threatened to rouse anti-British feeling across the Muslim world, much of which still lay within the British Empire. The British were themselves deeply divided about which party to support; targeted by Palestinian Arab and Jewish insurgents in turn, they attempted to reconcile the conflicts inherent in the contradictory promises they had made. Mandatory Palestine expired in open warfare. The British exit in May 1948 was more ignominious and scarcely less bloody than the Partition of India a few months before.

Our map shows land purchased by the Jewish National Fund (JNF), the Palestine Jewish Colonisation Association (PICA) and 'other' Jewish land. The issue of land ownership was central to Survey of Palestine mapping activity. These were cadastral rather than topographical maps, created for the purposes of taxation and land registration: 'the way chosen by the Mandate government to organise its commitments was the early organisation of the land system on a legal system for land ownership'.[61] The fairness of the decisions made by the survey teams remains controversial, but the scholarly view is that the Survey benefited from 'the best the Empire had to offer'.[62]

Jewish land ownership is better documented than that of the Palestinian Arabs. The purpose of private organisations such as the JNF and PICA, whose activities are indicated on the map, was to purchase land for Jewish use. The Jewish population grew steadily from a relatively low base, perhaps as little as 10 per cent of the total population in 1920 to around one third in the 1930s. The purchase of land, from which the existing Arab agricultural workforce was often excluded, was a highly visible activity that heightened tensions between the communities. Hope Simpson was concerned by the acquisition of the land marked on our map, which he regarded as 'extra territorial',[63] and he argued that Jewish immigration had to be regulated according to Palestine's economic capacity to absorb more people. The quota system that emerged from his report pleased no one but was applied for the remainder of the Mandate, against a background of Nazi persecution. From 1933 onwards Jewish refugees made increasingly desperate attempts to enter Palestine illegally, but the majority were intercepted, and interned or deported, even after the war.

1931: A map of New Zealand, Britain's outlying farm

CCORDING TO THIS large pictorial map, the most important element of New Zealand is its capacity to produce. New Zealand, the map informs us, is nothing short of an economic powerhouse, capable of feeding, clothing and powering an empire. The map is certainly persuasive, showing the two islands teeming with activity, focusing almost entirely on fish and sheep. Tables provide figures of livestock, crops, fish and minerals. But, above all, it is the stylised, bold and confident artwork that reinforces the positivity and credibility of image and message.

These are hallmarks of the work of the Empire Marketing Board (1926–33), a short-lived but prominent body set up to encourage consumers to buy British Empire goods, thus promoting trade between Britain and her dominions. Many varied posters and a great deal of promotional literature were produced by the organisation, with the purpose of maintaining economic bonds during years of world economic depression which tested those bonds to the limit.

Following the United States' economic crash and world depression, governments adopted protectionist policies to safeguard their currencies from inflation (see the 'New Treasure Island Chart', p. 82). Such policies, however, stifled trade by making tax on goods expensive for the importer, whose government generally returned the favour in their own policies. British politicians' attempts to solve this crisis included the suggestion of empire free trade – in effect, the establishment of an empire trade zone, which would have provided more amenable trade conditions within the British Empire. But it never happened. Places such as Canada and New Zealand desperately needed to protect their own industries, even from Britain. In New Zealand wool prices fell by 60 per cent in two years.[64] An Imperial Economic Conference at Ottawa in 1932 yielded only a series of individual agreements.

Up to 1933 the Empire Marketing Board had played the role of promoting and facilitating economic cooperation. One of their regular artists, MacDonald Gill, who produced many posters for other clients including the London Underground, the United Nations and the International Tea Market Expansion Board, created a world map for the EMB entitled 'Highways of Empire'. He also designed the board's logo and this map of Britain's 'outlying farm'.[65] Such images were not only practical tools for easing economic concessions between the dominions; they served a wider role in maintaining Britain's empirical ties, propaganda supporting the role of the Royal Navy (see the 1901 Navy League map, p. 20). The map's message was straightforward (as one would expect from an advertising poster) and deeply colonial.

There were other ways of maintaining bonds. In the same year as Gill's celebratory poster, the Conservative Prime Minister Arthur Balfour's definition of the Commonwealth as embodying a free association between members was enshrined in the Statute of Westminster. Far from being a dominion on the opposite side of the globe to the British Isles, this 'outlying farm' could be anywhere. However, any positive communication, by means of royal visits as much as cartography (see the 1953–4 royal tour, p. 128), was continually encouraged.

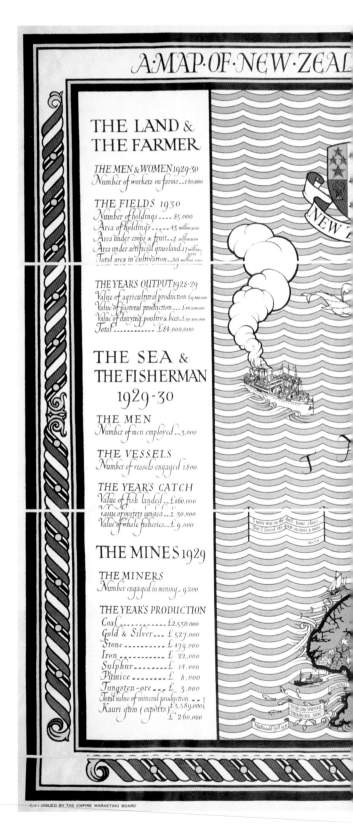

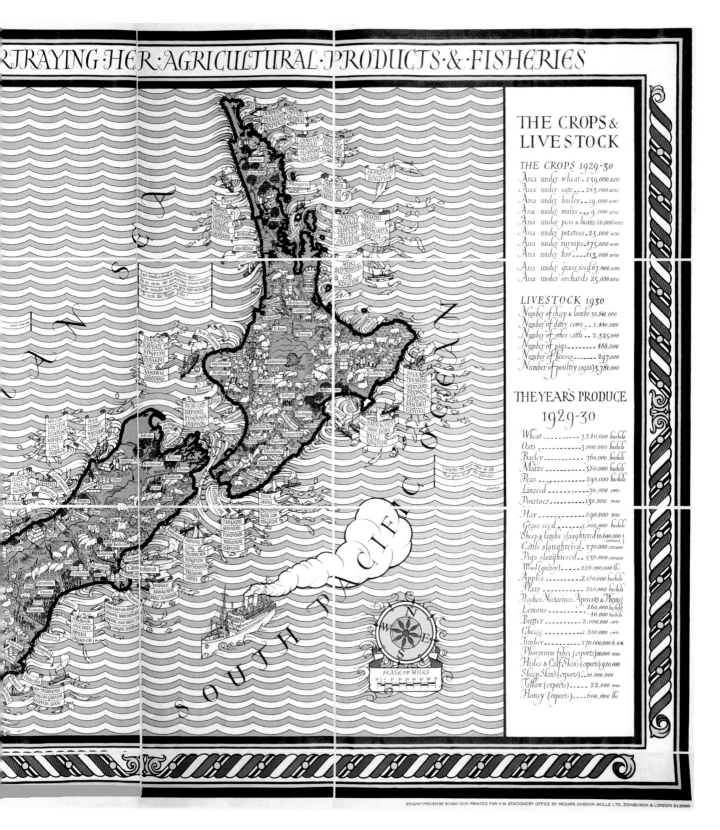

THE CROPS & LIVESTOCK

THE CROPS 1929-30

Area under wheat _ 239,000 acres
Area under oats _ _ 283,000 acres
Area under barley _ 19,000 acres
Area under maize _ _ 9,000 acres
Area under peas & beans 10,000 acres
Area under potatoes _ 23,000 acres
Area under turnips 475,000 acres
Area under hay _ _ 413,000 acres

Area under grass seed 67,000 acres
Area under orchards 25,000 acres

LIVESTOCK 1930

Number of sheep & lambs 30,841,000
Number of dairy cows _ _ 1,440,000
Number of other cattle _ 2,325,000
Number of pigs _ _ _ _ _ 488,000
Number of horses _ _ _ _ 297,000
Number of poultry (1926) 3,781,000

THE YEAR'S PRODUCE 1929-30

Wheat _ _ _ _ _ _ _ 7,240,000 bushels
Oats _ _ _ _ _ _ _ 3,000,000 bushels
Barley _ _ _ _ _ _ _ 760,000 bushels
Maize _ _ _ _ _ _ _ 380,000 bushels
Peas _ _ _ _ _ _ _ 290,000 bushels
Linseed _ _ _ _ _ _ 70,000 cwts
Potatoes _ _ _ _ _ _ 130,000 tons

Hay _ _ _ _ _ _ _ 890,000 tons
Grass seed _ _ _ _ _ 1,000,000 bushels
Sheep & lambs slaughtered 10,640,000 carcases
Cattle slaughtered _ _ 370,000 carcases
Pigs slaughtered _ _ 530,000 carcases
Wool (greasy) _ _ _ 226,000,000 lbs.
Apples _ _ _ _ _ _ 2,180,000 bushels
Pears _ _ _ _ _ _ 220,000 bushels
Peaches, Nectarines, Apricots & Plums } 260,000 bushels
Lemons _ _ _ _ _ _ 40,000 bushels
Butter _ _ _ _ _ _ 2,000,000 cwts
Cheese _ _ _ _ _ _ 1,800,000 cwts
Timber _ _ _ _ _ _ 270,000,000 ft. B.M.
Phormium fibre (exports) 10,000 tons
Hides & Calf Skins (exports) 920,000
Sheep Skins (exports) _ 10,000,000
Tallow (exports) _ _ _ 22,000 tons
Honey (exports) _ _ _ 600,000 lbs.

251Q/WT.P1013/1139 30,000 10/31 PRINTED FOR H.M. STATIONERY OFFICE BY MESSRS DOBSON MOLLE LTD., EDINBURGH & LONDON 51.2590

THE NEW NORMAN CONQUEST

1933: Pieces of eight: a satirical cartoon of the world financial crisis

Y THE TIME of the 1929 American stock market crash and the crippling of the world economy, Britain had already endured years of economic stagnation and rising unemployment. The real crisis came in September 1931, when its inability to balance payments and the resultant drain in confidence provoked a run on its currency, the pound. The Labour government fell and was replaced by a national coalition. Among the measures introduced, the abandonment of the gold standard was the most controversial: this safe, protectionist, anti-inflationary policy (whereby only as much money was produced as existed in gold reserves) had been the peacetime bedrock of British economic policy for a century.

Unlike in previous eras, a well-informed, literate and enfranchised public felt entitled to know the business of the Treasury and the Bank of England. Over the 1920s, newspaper readership had increased dramatically, and by 1933 the *Daily Express* boasted 2 million daily readers. *The Express* and its rival *Daily Mail* espoused popular, accessible and sometimes aggressive journalism, and were enlivened by photographs, illustrations and cartoons. It is in such newspapers that we occasionally see the map used as a satirical device. In the *Express* of Monday 23 January 1933, the cartoonist Sidney 'George' Strube's 'A Chart of the Financial Main' was published.

Evoking the famous map from Stevenson's then fifty-year-old novel *Treasure Island*, the map presented a strong argument against motions by the Bank of England to rejoin the gold standard under pressure from the United States in 1933, by citing examples of wasted British taxpayers' money and incidents from the previous years. The years after 1931 had not been dramatically better than those before. Unemployment in 1929 had been 8 per cent, but by 1932 the figure had risen to 17 per cent, the worst level since records began.

In fact, the battleground was not between currencies (or their nations), as presented in the cartoon, but between the ideologies of protectionism and free trade. On one side the suave pirate Long John

Skinner, persuading the bowler-hatted 'little man' (a regular figure in Strube's cartoons) to sign back on to the gold standard, was the Governor of the Bank of England, Montagu Norman. Nicknamed Professor Skinner after an incident of mistaken identity by the American press (he used aliases when travelling), it was Norman who had returned Britain to the gold standard in 1926 after the economic strain of war had necessitated leaving it. On the other side, not illustrated, was Max Aitken, Lord Beaverbrook, Canadian millionaire *Daily Express* owner and vocal opponent of protectionism, who had fought a local election in 1931 from the position of empire free trade. Trade agreements with the dominions meant not dealing with rival currencies the dollar and the French franc.

In Strube's cartoon, the solidarity between these wartime allies is gone, the physical distance between them magnified dramatically. The 'land' massed around the Atlantic is similarly distorted, yielding the profiles of Uncle Sam and a moustachioed Frenchman (represented by her 'Gold' Coast colony). British grievances include resentment at unfair practices by the French and Americans in amassing vast stockpiles of gold, as well as distrust, xenophobia, and bad blood over war debt and reparations – money wasted that could have been put to 'good' use. This would have made an impression on the conscientious reader, who would also have read about, or even seen, the hunger marches of 1930 to 1933.

Nobody need have worried: five months later President Roosevelt took the USA off the gold standard too. What at first may appear unthinkable may become normal very quickly. The day after Strube's cartoon, which had no doubt been feverishly worked up over a weekend with none of the time and consideration that even the 'Hark! Hark!' illustrator (see p. 46) may have enjoyed, another cartoon carrying another point went to press. The son of a German-born wine dealer, Strube became one of the most popular political cartoonists of the 1930s; his anti-Nazi cartoons earned him a place on the hit list prepared by the Gestapo for the invasion of Britain.

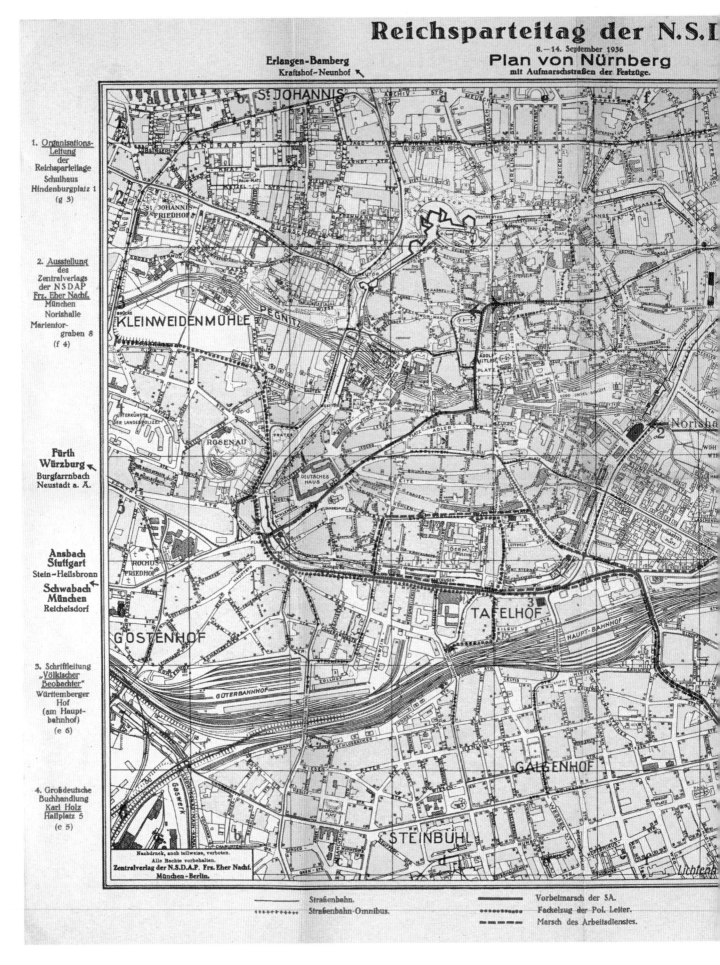

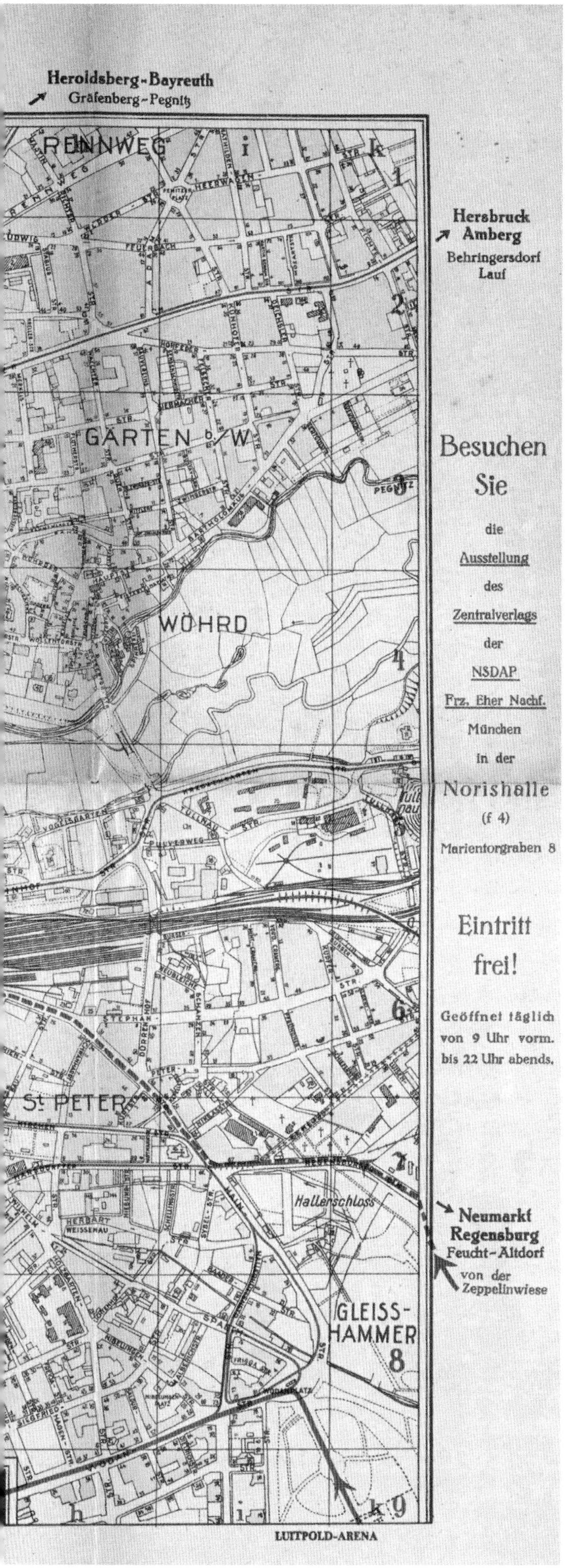
Heroldsberg - Bayreuth
Gräfenberg - Pegnitz
RENNWEG
Hersbruck
Amberg
Behringersdorf
Lauf
GÄRTEN b/W
WÖHRD
Besuchen Sie die Ausstellung des Zentralverlags der NSDAP Frz. Eher Nachf. München in der Norishalle (f 4) Marientorgraben 8
Eintritt frei!
Geöffnet täglich von 9 Uhr vorm. bis 22 Uhr abends.
St PETER
Hallerschloss
Neumarkt
Regensburg
Feucht - Altdorf
von der Zeppelinwiese
GLEISS-HAMMER
LUITPOLD-ARENA

1936: Hitler's 'best *Parteitag*': Nuremberg and fascist tourism

THE NUREMBERG RALLIES became one of the sights of 1930s 'fascist tourism', attracting various wealthy and well-connected Britons including repeat attenders Diana and Unity Mitford.[66] Hitler valued their presence, granting the aristocratic sisters remarkable access to his inner circle. In Diana's letter to Unity of 17 September we glimpse the 1936 Reichsparteitag through British eyes:

> I must tell you how *sweet* the Führer was. He came into the room and made his beloved *surprised* face, and then he patted my hand ... he was so wonderful and really seemed pleased we had gone every day ... I said we loved the wonderful parades and he said it was the best *Parteitag* he had ever had because *everything* had *geklappt* [worked].[67]

The Mitfords would have seen (if not used) official maps such as this one. Spare, functional, unencumbered with party slogans and insignia, maps played an essential role in one of the most meticulously orchestrated acts of political theatre of the twentieth century, guiding hundreds of thousands of party functionaries, diplomats and journalists who descended annually on the ancient city of Nuremberg for the Nazi celebration of the power and unity of the Third Reich. The routes picked out in red on the map provided a framework for the set dressing: the banners, searchlights and torchlit parades, still familiar today from the footage shot by Leni Riefenstahl and others, symbolising the public face of National Socialist Germany.

Nuremberg had no particular associations with the earliest years of the Nazi movement and was not especially noted for Nazi sympathies. It was chosen for the first rally in 1927 partly because it was a historic German city – which suited Nazi ideology – but principally because of its convenient central location. Another rally went ahead in 1929, but the town authorities blocked repeat performances in 1930 and 1931, refusing to make suitable venues available. Even after Hitler came to power, he had considerable difficulty in obtaining planning permission for the monumental complex which he envisaged as a suitable backdrop. He got his way on commercial rather than ideological grounds: the rallies brought in a massive amount of business.

Implications for residents – which may explain local scepticism – are there on the map. What remains, essentially, a medieval street plan is dwarfed by the brutal new complex. Parades through Nuremberg's narrow streets suited the romantic, *völkisch* visions of Germany's past that the Nazis were so keen to foster, but daily life ground to a standstill. The scale of the vast new stadium complex outside the city was in proportion to the rallies rather than Nuremberg itself; tens of thousands had to be accommodated in temporary camps around the city, where there were incidents of drunkenness and vandalism. Hitler's followers were in an especially celebratory mood when they congregated at Nuremberg in 1933, following his appointment as Chancellor, and after the rally the party was faced with a huge clear-up bill for unpaid drinks, stolen beer mugs and cleaning graffiti from public transport; the toilets were left in a particularly foul state. Orders distributed in advance of the 1934 rally were categorical: 'One thing holds true above all else for this occasion: a National Socialist does not get drunk'.[68]

Over the course of the 1930s the rallies became more professional but ever more bloated, doubling in length from four to eight days. With an eye on his successors, Hitler also set about the canonisation of what could be termed the liturgy of the event; there was an identifiable underlying quasi-religious element, with Hitler himself officiating over an emotionally charged, communal experience in the role of 'high priest'.[69] Whatever Hitler's plans were for a Thousand Year Reich, evidence suggests that by the end of the 1930s elements of the party rank and file were nevertheless already bored by the endless assemblies and marches.

1936 was a key year. The Berlin Olympics had run smoothly, without any serious boycotts, and the international community had accepted both the deployment of German troops abroad (the 'volunteers' of the Condor Legion) and the re-militarisation of the Rhineland, carried out in flagrant disregard of the 1925 Treaty of Locarno. The mottoes of each rally reflected the situation at the time (the cynical 'peace' rally of 1939 was scrapped on the outbreak of war); the 1936 rally was the rally of 'honour' – reflecting both the restoration of honour to the German people and international acceptance of the Nazi regime. Although Germany was spending far more than either Britain or France on rearmament, only Hitler and his generals knew how weak the Wehrmacht still was, and how ready they had been to abandon the Rhineland again at the first sign of decisive action by the other powers. The rally of 1936, displaying the supposed new military might of Germany for all to see – abroad as well as at home – marked a shift in policy, from concealment to intimidation and bluff. The stage was set for appeasement and a series of effortless victories for Hitler between 1936 and 1939.

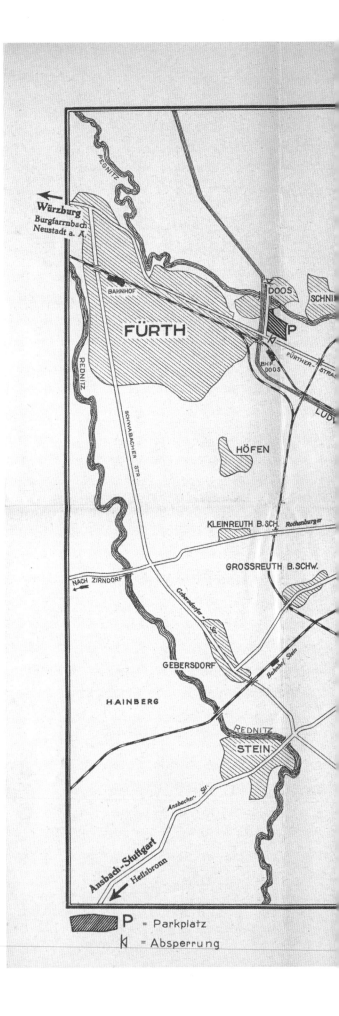

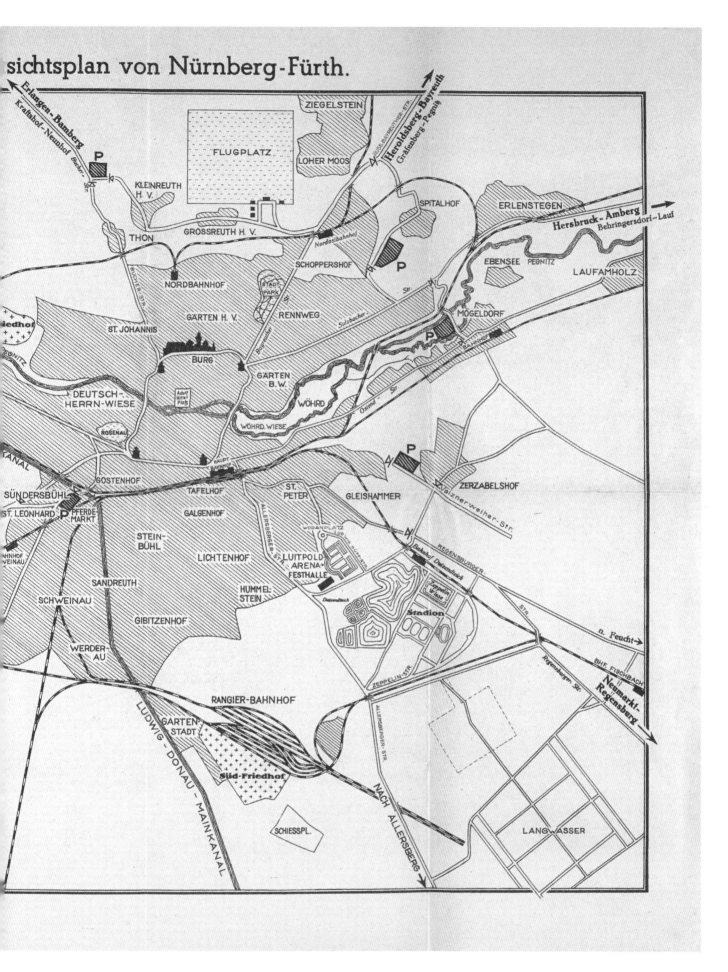

sichtsplan von Nürnberg-Fürth.

1937: A souvenir flag of the coronation of Edward VIII

THIS UNION FLAG is one of a great number mass-produced in 1936 for the coronation of King Edward VIII (1894–1972). Edward had acceded to the throne upon the death of his father, George V, on 20 January 1936, with his state coronation planned for 12 May the following year. Printed on to fabric, the flag has at its centre a half-portrait of the king in between a twin hemisphere world map showing the empire, his dominions, in red. But the coronation of Edward VIII never happened. The king abdicated on 11 December 1936 after less than a year on the British throne, under pressure from a government fearful of public reaction to his plan to marry an American commoner called Wallis Simpson. Mrs Simpson was a divorcee and, as such, was not eligible to be married in a Church of England service, the church of which the king was head.

Edward VIII, it seems, chose love over duty, and for the substantial lobby who would have preferred to keep this erratic character with suspected fascist sympathies out of the way, it was a godsend. Yet among souvenir manufacturers the supposed romance of his decision may not have warmed many hearts. Due to the voluntary secrecy of the British (though not the international) press over the affair, the earliest the public knew of it would have been in October. Flags, along with countless coronation mugs, plates, stationery, images, stamps and medals, became superfluous virtually overnight. However, for these objects, their makers, and everyone else besides, the abdication wasn't the end of the world. Edward's brother was crowned George VI on the same day prepared for in May and the outdated souvenirs for Edward's coronation have today achieved a particular value (though not a high one) as curious collectables.

Looking at the flag and its constituent images of patriotism, it is easy to project on to it a symbolism that presciently matches a fallen king with a crumbling empire. It is doubtful that many, if any, originally saw in it such a thing. The dominions in particular – nations who had fought as empire troops during the First World War – identified very strongly with British patriotism and its royal family. Many identified with the king, who was handsome, charismatic, charming and extremely popular in the dominions. His tour of Canada had been a huge success (he bought a house in Alberta in 1922), and at the end of 1916 he visited Anzac troops freshly evacuated from the Gallipoli Peninsula (see p. 50).

A more accurate reading of the flag is that the positioning of the British monarch at the centre of the hemispheres demonstrates his crucial importance as a lynchpin of the empire. That the pin might become unstuck was certainly the fear of Prime Minister Stanley Baldwin. Consulting with the governments of dominions, and feeling himself a good judge of public opinion (over Churchill and Lord Beaverbrook, who felt the king should have stayed), he doubtless felt, along with many others, that the symbolic power of the monarchy was too important to the togetherness of the British Empire to be put in the hands of Edward VIII. The identity of the king in the portrait was of little consequence to the function of the flag as a patriotic symbol, and of far less importance than the affirming, loyal red of each and every empire dominion.

1938: The Spanish Civil War: British neutrality and poor-quality maps

UNLIKE OTHER EUROPEAN signatories of the non-intervention agreement, Great Britain and France remained neutral in the Spanish Civil War (1936–9). But despite the British government's position, which became more controversial with every reported atrocity against Spain's civilian population, it retained interest in the course of the war. By the end of 1936 the conflict had widened from a Nationalist military coup against the fragile Republican government into a proxy world war. Germany and Italy supplied troops and weapons to the Nationalist rebels under General Francisco Franco, Russians supplied somewhat less to the defenders. Britain could observe her enemies in action from a distance, but she also had something to lose, being acutely aware that British naval interests in the Mediterranean Sea rested with Gibraltar (as well as Suez, see p. 134). Furthermore, some two thousand British citizens, mostly local Communist Party members, were in Spain, fighting voluntarily for the 'International Brigade' against fascism.

Restricting Britain's gaze was the poor quality of military and topographical mapping of Spain. Not since the Peninsular war of 1808–9 had Britain (or anyone else for that matter) been militarily engaged in Iberia.[70] This dearth of intelligence material provided just one of many reasons for non-intervention. Crude, derivative 'operations room' maps were produced to illustrate particular battles, including that of Malaga (1937) and the blockade of Bilbao (also 1937, watched by British warships in the Bay of Biscay). But it was only in 1940 that a large-scale topographical map of Spain was accessioned by the War Office and reprinted. The date of the original was 1886.

Map coverage of Spain was not a priority, but any map was better than no map, particularly if the interest was merely that of a bystander. Decades after the end of the war, in 1962, a large lithographic map of Spain was added to the map room inventory of the Directorate of Military Survey in Tolworth. Crudely printed, yet impressive in detail, with lines of communication, the map had been printed in Bilbao in the summer of 1938. It is a Nationalist propaganda map, bearing slogans such as 'Saludo a Franco'. Colour defines the territorial gains made by the Nationalist army as stages in the conflict. The eastern part of Spain, still in Loyalist hands, is uncoloured, or rather yet to be coloured. Then reaching its final stages, the siege of Madrid had been portrayed with unusual force by a British press unable to decide which out of fascism and communism was the greater evil.

By 1962 the ideological struggle between Right and Left was reaching another peak with the Cuban missile crisis, and the Spanish Civil War was history. Spain enjoyed prosperity, the tourist boom attracting a large number of British holidaymakers to Spain every summer (see the tourist map of Alicante, p. 138). The Ministry of Defence was happy to continue to rely on ephemeral and outdated maps. During the Civil War some British Intelligence officers had been permitted to observe military tactics on the front, as guests of Franco's army. Their unofficial position was generally in favour of a more stable Nationalist Spain.[71] The Nationalist map may well have been acquired at that time and only stamped with an accession date when it was uncovered in the map room, possibly during a search for Spanish mapping, decades later.

1940: Blitzed London: a Luftwaffe map of 'Mayfair Square' and the London County Council 'bomb damage' maps

FROM 7 SEPTEMBER 1940 over three thousand sorties, mostly at night, were flown by the Luftwaffe against London. This was the Blitz, the aerial bombing campaign upon London and other industrial and cultural British urban centres, which reduced many to rubble.

Here we compare two sets of maps, both based on the Ordnance Survey. The map of 'Mayfair Square' is an extract from Ordnance Survey's 6-inch map of London, originally published in 1921. This example was reprinted for the Luftwaffe: marked 'secret' ('geheim'), labelled 'Lft. Kdo. 2' (for the squadrons of Luftflotte 2) and dated September 1940, it was used by the unit that carried out the first raids, by day and night, in what became one of the most terrifying and destructive episodes in London's history. Throughout that campaign, the 1937 Ordnance Survey 5-foot scale maps of Greater London were hand-coloured by staff in the Architect's Department of London County Council to record the damage, building by building, ranging from superficial to total destruction. Updates from 1944 onwards included the impact sites of V-1 and V-2 rockets. After the war the maps were used to inform demolition and reconstruction, and they are still in demand today from surveyors: subsidence in modern London is not always caused by tree roots.

The Luftfahrtministerium made great use of Ordnance Survey mapping in identifying strategic military and utility targets, reproducing them in *Sonderfolgen* (special instruction manuals) for the use of bombing crews. Having been photographically transferred, our map was overprinted in red to provide data and target outlines. Alongside aerial photographs and instructions, copies would have been used for training and possibly for navigation on board one of the hundreds of bombers flying from bases in Belgium and France. The map marks what to avoid, as well as what to destroy, and makes use of landmarks for tracking – such as the distinctive shape of Broad Street, below New Oxford Street. It reflects optimistic early assessments of the level of precision bombing that was possible, even by daylight.

From August 1940, after the Dunkirk evacuation, the Luftwaffe attempted to nullify the threat of the Royal Air Force in order to achieve the air superiority necessary for a German amphibious invasion of England (Operation Sea Lion), first by destroying British radar stations and then Fighter Command's airfields. On both occasions the Luftwaffe changed tactics when on the brink of success. The switch in September from RAF facilities to urban strategic utility targets, such as power and pumping stations, gasworks, railway depots and 'choke points', was designed to distress and demoralise the civilian population. Destructive though it was, it cost the Luftwaffe the battle, if not the war.

The identity of the Mayfair targets reveals that in September 1940 this switch had still not taken place altogether. The War Office and Admiralty, together with Somerset House (which contained some Admiralty staff), are joined by the large building to the south-east of Berkeley Square. This is the vast Air Ministry Building, which was only completed in 1938 and thus not present on the 1937 LCC Bomb Damage map. The building on the Luftwaffe map is also the earlier, demolished structure.

A large area of Belgravia populated by embassies, including that of the United States, is labelled Neutrale Botschaften. Bombing here would not have been diplomatic – there was still every possibility that a strong isolationist lobby might keep America out of the war – yet the bomb damage map section shows that over the following four years this area sustained more damage than any of the targets of 'Mayfair Square' except the War Office and Admiralty.

Within a month daylight raids had all but ceased: Luftwaffe losses proved unsustainable. However, Britain's night air defences also improved rapidly. As airborne radar fitted to night-fighters and ground-based radar directing anti-aircraft guns became more effective, bombers were forced to abandon individual targets in favour of area bombing. After November 1940 precision bombing was accompanied by the tactic of terror bombing upon civilian targets, which had by its nature even less need of specific red overprinting.

Operation Sea Lion was postponed indefinitely in mid-September. British cities suffered a terrible punishment after the Luftwaffe's change of strategy, but they absorbed it. Britain would become the embarkation point for the invasion of Hitler's Fortress Europe, and lessons learned in the Blitz would be devastatingly applied – with technology and tactics advanced by years of war – to German cities in 1944–5.

1940: Occupied Paris: a tourist map for German troops

THIS GERMAN-LANGUAGE TOURIST map of Paris was presented with the compliments of the military governor (Kommandant) of occupied Paris to German troops on leave. It is dated October 1940, a matter of months after the fall of France and Hitler's own sightseeing tour of the city, when he posed for photographers in front of the Eiffel Tower.

The contrast with the two preceding maps of blitzed London could hardly be greater. The speed of the German victory surprised everyone, including the Germans. Paris was declared an open city by the French government on 10 June 1940, and German troops entered Paris unopposed on 14 June. Last-ditch negotiations to continue the war via a full Franco-British union were rejected (Marshal Pétain, about to assume the role of head of the Vichy state, likened the proposals to being chained to a corpse) and an armistice with Germany was signed on 22 June. Central Paris was not bombed by the Allies as it was not an industrial centre, and the last Kommandant allegedly ignored Hitler's orders to destroy the historic city as the Germans retreated. The fabric of Paris survived the war largely unscathed.

The complexity of the relationship between the French people, the Vichy regime, the occupying forces and their erstwhile allies, the British, is explored more fully on p. 100 (a Vichy propaganda poster). However, in the context of this map it is worth noting that the nature of the German presence was very different to that in other occupied territories, especially in Eastern Europe. A German soldier in Paris could savour victory, but also escape the horrors of the war elsewhere.

This is reflected in our map. It is an economical piece of two-colour printing, produced locally by Parisian firm Mouillier & Dermont. It appears to be newly commissioned, which is significant as it is a simple tourist map. No sites associated with the occupation are marked: the buildings shown in relief are, more or less, the traditional attractions. The ornate gothic script is a sufficient statement of control.

The same holds true of the guide stapled to the cover of the map. There is no résumé of rules and regulations concerning fraternisation, or how the civilian population should be treated. That is not to say that the selection of places of interest is entirely without nuance. The Parisian version of the Lion of Belfort commemorates a French victory in the Franco-Prussian War. The children and grandchildren of veterans of that conflict were serving in the German military of 1940. The inclusion of the monument may have been inspired by sentiments similar to those that led Hitler to choose, for the site of the French surrender of 1940, the same railway carriage near Compiègne where the Armistice of 1918 had been signed. The defeats of the past had been expunged by the decisive victory in the present.

The Paris Mosque was also singled out as a place of interest. Completed in the 1920s, a tribute to the many thousands of Muslim troops in the French army killed during the First World War, it was a relatively new building. Nevertheless, the guide extols the mosaics and the fine carvings in Moroccan cedarwood. We have explored the German relationship with the Islamic world earlier in the twentieth century on p. 52 (the Berlin–Baghdad railway) and the inclusion of this mosque here could be seen as an extension of that sentiment. However, the Paris Mosque ultimately served as a refuge for Jewish people fleeing the Holocaust. The influential imam of the mosque, Si Kaddour Benghabrit, saved as many as a hundred by providing them with certificates of Muslim identity.

Pariser Plan

LEVALLOIS

NEUILLY

PANTIN

LE PRÉ
SAINT-GERVAIS

BOIS DE BOULOGNE

BOIS
DE
VINCENNES

CHARENTON

ISSY

VANVES

MONTROUGE

IVRY

Das
Geschichtliche
und
Kunsthistorische
~ Paris ~

1. — Louvre Palast.
2. — Triumphbogen Carrousel.
3. — Pl. de la Concorde - Obelisk.
4. — Grand Palais.
5. — Etoile. Triumphbogen mit dem Grab des Unbekannten Soldaten.
6. — Chaillot Palast.
7. — Eiffelturm.
8. — Militär Schule.
9. — Dôme des Invalides.
10. — Abgeordneten Haus (französische Kammer).
11. — St-Germain-des-Prés Kirche.
12. — Französische Akademie.
13. — Luxembourg Palast (Senat).
14. — Cluny.
15. — Panthéon.
16. — Moschee.
17. — Notre-Dame.
18. — Hl. Kapelle.
19. — St-Jakobsturm.
20. — Rathaus.
21. — Vendôme. Platz mit Säule.
22. — Hl. Magdalena Kirche.
23. — Opernhaus.
24. — Börse.
25. — Augustinus Kirche.
26. — Hl. Herz Jesu Kirche.
27. — Tor Saint-Denis.
28. — Tor Saint-Martin.
29. — Statue der Republique.
30. — Bastille Platz mit Gedenksäule.
31. — Nation Platz.
32. — Park Buttes-Chaumont.
33. — Loewe von Belfort.

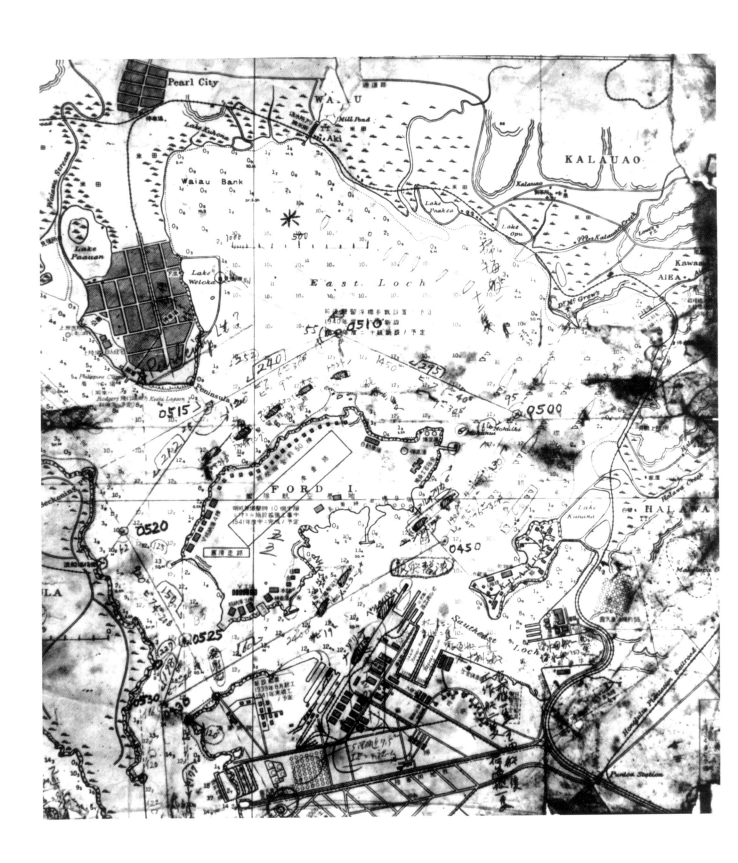

1941: A spy map for the Japanese submarine attack upon Pearl Harbor

THIS MAP WAS apparently discovered in one of the Japanese Ko-Hyoteki 'midget' submarines that attacked Pearl Harbor on 7 December 1941. It is a map of the Hawaiian harbour, probably derived from a United States survey chart, overwritten in Japanese and extensively annotated to assist the navigation of the submarines and to identify targets. Straight lines and angles mark, with timings, the paths of Japanese aircraft attacking from above. The submarines were to enter the harbour from bottom left, then travel anticlockwise around Ford Island past 'Battleship Row', where the might of the American Navy was then moored. Instructions written on it include 'sink an American ship'.

The surprise attack upon the US Pacific naval base by Japanese torpedo planes, bombers and submarines damaged or destroyed seven battleships and killed over two thousand people. It paralysed the US Navy, though only temporarily. The extent of this 'surprise' has been the subject of debate for over half a century. Although worsening Japanese–American relations meant that an attack was likely, there were many possible targets besides Pearl Harbor. For the United States, the presence of her fleet in Pearl Harbor was a threat to Japan in the Pacific; to the imperial Japanese Air Force, the fleet was a sitting duck.

Despite the apparent inadequacy of the harbour's defences, which Lord Mountbatten, admiral of the British fleet, had noted during a visit earlier in 1941, a Japanese attack required information on what they were attacking. Maps of military bases have historically been closely guarded by their occupants. The paucity of maps in the British Library of Pearl Harbor after 1898 (when the peaceful lagoon harbour first became a US naval base) attests to that.[72] Much of the incredibly precise intelligence included on the Pearl Harbor map may have been supplied by a single Japanese spy in Honolulu named Takeo Yoshikawa, who compiled weekly 'ships in harbour' reports.

In the event, this map was not required since the midget submarine in which it was found ran aground outside the harbour and was captured. But what it showed that Japan knew of Pearl Harbor caused acute embarrassment to the US military, and it mysteriously disappeared from FBI custody, apparently 'purloined by a souvenir hunter'.[73] The map is known today only through a photograph in the Naval Aviation Museum in Pensacola, Florida, which crops the image on at least one side.

After capture, the submarine and its surviving operator Kazuo Sakamaki (the first Japanese prisoner of war) were displayed before the US public as evidence of aggression and provocation. The US declared war on Japan, and the fellow signatories of Japan's 1940 Tripartite Pact, Italy and Germany, declared war on the US. Britain, exhausted and isolated as it was by late 1941, was thrown a lifeline.

1942: Vichy and the Churchillian octopus

THIS BLOATED CHURCHILLIAN octopus represents the official view from Vichy France of Britain's decision to carry on the war alone. Its anti-English sentiment does not automatically mean that it would have been regarded (or reviled) by its intended French audience as Nazi propaganda – the nature of support for the Vichy regime was more complex than that.

A tradition of cartographic cephalopods stretches back to Fred W. Rose's 'serio-comic war map' of 1877, satirising the Great Eastern Crisis. The octopus serves to dehumanise the enemy, and its tentacles are a convenient and sinister metaphor for territorial ambitions on an international scale, reaching across borders or oceans to choke the life from some other hapless nation. This poster urges the reader to have faith – the tentacles are being severed 'methodically'. Healthy tentacles still reach out towards North America, South Africa, Aden and India, but elsewhere they have been bloodily truncated by the Axis powers. Norway, Libya and British Somaliland refer to German and Italian campaigns, but the other names would have resonated directly with a French audience. These do not necessarily commemorate Vichy French victories, but insofar as the loss of French lives made recruiting for de Gaulle's Free French even more difficult, these events of 1940–1 could certainly be regarded as wounds on the octopus.

Taking these wounds one at a time, the refusal to scuttle or surrender the mighty French fleet at Mers-el-Kebir in French Algeria, or sail it to a neutral American port, culminated in a bombardment by the Royal Navy that killed 1,297 French sailors, causing great shock and resentment. The French felt perfectly capable of keeping their fleet out of German hands, and they needed it to protect their colonies; the British decided that they could not risk the French losing control. The raid on Dakar was a fiasco, memorably fictionalised in *Men at Arms* (1952) by novelist Evelyn Waugh (who was there). In 1941 British, Indian, Australian and Free French troops invaded Syria to secure the Allied flank, particularly after the Luftwaffe had used Syrian bases to attack Iraq. Vichy resistance was stiff, stoked by old rivalries in the Middle East and with an eye to the post-war balance of power in the region.

The viciousness of the poster may surprise the modern reader, but that would be a misreading of contemporary French politics. While many resisted the German occupation, support for the Vichy regime was far more widespread than it was expedient to admit after 1945.[74] As our unidentified artist 'S. P. K.' makes no reference to the British invasion of Vichy-controlled Madagascar, it seems likely that the image was created in the first part of 1942. At that time the agreement negotiated by Marshal Pétain's government after the fall of France in 1940 still held. Pétain, the hero of Verdun, was the last premier of the Third Republic and his new administration, based in the spa town of Vichy, held nominal authority over the whole of France. In the north, the 'occupied zone', this was very much at the discretion of the German forces, but the south remained a free zone until November 1942 and France's colonies were also left in Vichy hands. Vichy represented a significant measure of sovereignty.

Most of the French soldiers rescued from Narvik and the beaches at Dunkirk elected to be repatriated to France after the French surrender rather than join de Gaulle's Free French, and attempts to sequester French vessels in British ports were often fiercely resisted. It was a pattern repeated again and again where Allied troops came into conflict with Vichy forces. Some were committed to seeing France play a leading role in Hitler's New Europe, while others, separated from their families in France, simply wanted to play no further part in the war if they could avoid it. However, for the majority, it may simply have been that Pétain and his regime were seen as the legitimate government of France in contrast to de Gaulle, a lofty upstart without a mandate of any kind.

The campaigns and battles commemorated on the poster do not trip off the tongue: Mers-el-Kebir, Dakar and Syria were not extensively featured in British propaganda or celebrated since. There was little appetite for killing one's allies of a few months before, though there was anger at a perceived waste of lives. France came perilously close to civil war and in the immediate post-war era was keen to construct a myth of widespread national resistance. Our contemporary poster strips that away to reveal the divisions and uncertainties of the early 1940s.

1943: Blake's *Battle of the Atlantic*

THIS 1943 BRITISH propaganda poster, *The Battle of the Atlantic*, by Frederick Donald Blake (1908–97),[75] is most commonly encountered with English text. However, Blake's posters were part of a series produced for distribution abroad in various languages, including French, Dutch, Portuguese and – as here – Arabic, bringing the Allied message to the widest possible audience. This poster could easily have been destined for Egypt, Iraq, or any other region within the empire where Arabic was widely spoken. The text appears to be a faithful rendering of the English original and not tweaked or reworked for a particular audience, perhaps because the message is so very simple: the Allies have won the Battle of the Atlantic.

Britain is, effectively, Orwell's 'Airstrip One': nothing but factories, shipyards and gigantic concrete runways. Far from being enclosed by a U-boat ring of steel, waves of Allied aircraft radiate out. With air supremacy comes protection for the convoys steaming in from North America and those steaming out – the Arctic convoys bound for the USSR and convoys bound for the Mediterranean. In the mid-Atlantic U-boats are scattered and destroyed, and Hitler's Fortress Europe is under constant attack, with aircraft and parachute mines battering strategic targets such as railways, docks and submarine pens.

As propaganda, Blake's 1943 poster is not necessarily constrained by reality, but successful propaganda often manipulates a perceived truth and the Battle of the Atlantic really had turned decisively in the Allies' favour in the spring of that year. In March 1943 the U-boat 'wolf packs' came as close as they ever did to cutting Britain's Atlantic lifeline, and supplies of fuel and other vital resources reached critical levels. The situation was reversed within two months: Allied resources were freed from other theatres, and new long-range aircraft – which could now be fitted with sea-scanning radar and airborne depth charges – closed the mid-Atlantic gap. The wolf packs were harried out of existence and losses to Allied shipping were negligible in comparison with what had gone before. In May (dubbed 'Black May' by the U-boat crews) the Germans lost thirty-four U-boats in the Atlantic – an unsustainable one submarine for each Allied ship sunk. One lucky convoy (SC 130) escaped entirely unscathed, while between three and five of the attacking U-boats were destroyed. The German naval commander Karl Dönitz conceded defeat. As in the Great War a generation earlier, unrestricted submarine warfare failed to deliver a knockout blow before America entered the war, or before the build-up of men and materiel from America could prove decisive. One-sided as Blake's vision is, it still reflects the changed strategic situation.

Here is the English version of the text:

A ceaseless battle is raging in the Atlantic. The Axis U-boats' intention is to isolate and starve Britain. But as the U-boat offensive mounts so too do Britain's protective measures. More and more vessels are safeguarding convoys. The U-boat's Atlantic bases are being pounded by the Allied Air Forces and the entrances to their harbours are being mined from the air. The factories where they are built are being crippled by bombs. All these measures enabled Mr Churchill to say, when reviewing the U-boat campaign in May 1943: 'Our killings of the U-boats … greatly exceeded all previous experience and the last three months, and particularly the last three weeks, have yielded record results.'

1944: On leave in New York: a subway map for British sailors

THIS NEW YORK City Subway map is a 'special', commissioned by the Union Jack Club to show the location of its premises in New York. It was printed by George J. Nostrand in the style of the Hagstrom map, the official map of the system in the 1940s. Prior to 1940, New York's subway system had been operated by three private companies, each of which issued their own maps that gave priority to their own lines (in contrast to London, where independent companies engaged in joint branding from the Edwardian period onwards: see *How to Get There*, p. 36). Hagstrom's commercially produced map was the only one to give equal weighting throughout the system, and when the lines were merged under public ownership it was chosen as the official map.

Hagstrom disliked Beck's diagrammatic approach to the London Underground and, although there are some distortions, his map of New York City is more or less geographically accurate, retaining features of surface topography such as Central Park and the Statue of Liberty. Nostrand's map, with its distinctive red edges, is so similar that one can only speculate as to why Hagstrom appears to have taken no action against the rival publisher. However, while the origins of both maps remain obscure, it is impossible to be certain about who was copying whom.

This particular example was posted by the Union Jack Club to a British sailor in March 1944, care of the British Naval Liaison Officer at the Navy Yard, Charleston, South Carolina. The club had arranged accommodation with a hospitable American family for part of his leave, and the map was accompanied by travel and contact information so that he could reach his hosts in New Rochelle, just to the north of New York City.

The original Union Jack Club was opened in London in the aftermath of the Boer War, catering for both serving and former servicemen of non-commissioned rank. The New York branch was opened before Pearl Harbor brought America into the Second World War (see p. 98). It was based in a three-storey building rented from the prestigious Algonquin Hotel by the British War Relief Society – a US charity that coordinated vast fund-raising drives to provide non-military aid for Britain. The Duke of Windsor (suspected of pro-fascist sympathies and appointed Governor of the Bahamas in the hope that he could do the minimum damage to the war effort from the other side of the Atlantic: see p. 88) visited the club when he travelled to New York in October 1941. An early piece by the respected journalist Daniel Lang,

published in the *New Yorker* in September 1941, described the club as a place where seamen could 'drop in to drink beer, play darts, catch a snooze, or discuss girls, battles and the state of their commander's disposition'.

Lang expected anything between a dozen and twenty Royal Navy personnel to be present, sailors whose ships had put into neutral American ports for repair or refuelling. However, once America entered the war, thousands of uniformed servicemen and women could be seen on the streets (and in the bars, theatres and clubs) of New York – much as in London – many more of whom were from Britain and the Commonwealth. New York was a major port of embarkation for the European and Pacific theatres of war, but it was also a chance to sample the pleasures of civilian life. New York's entertainment industry enjoyed a boom period.

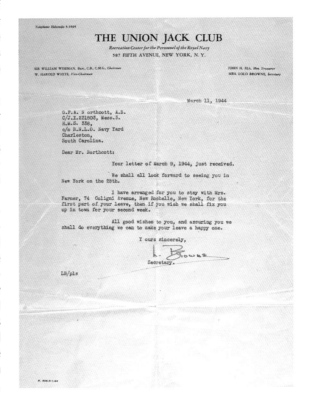

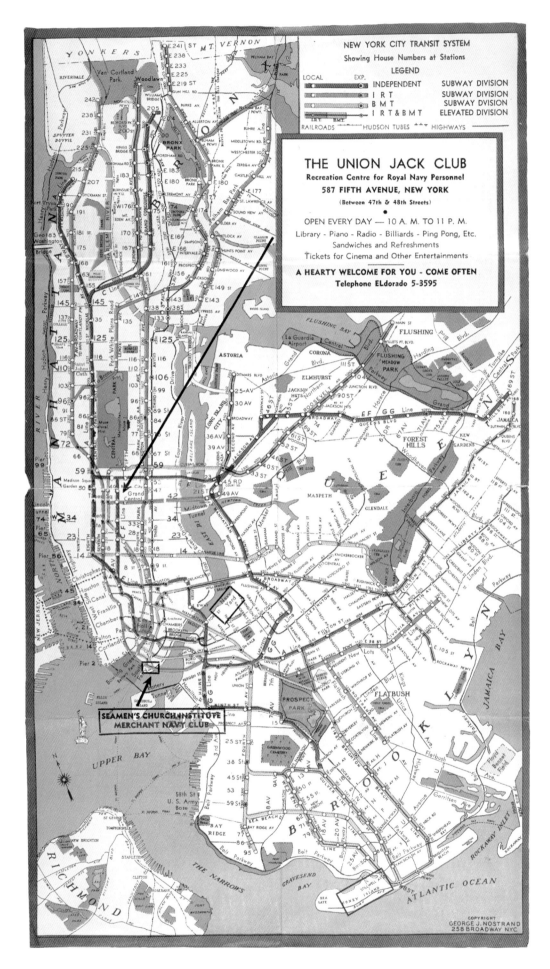

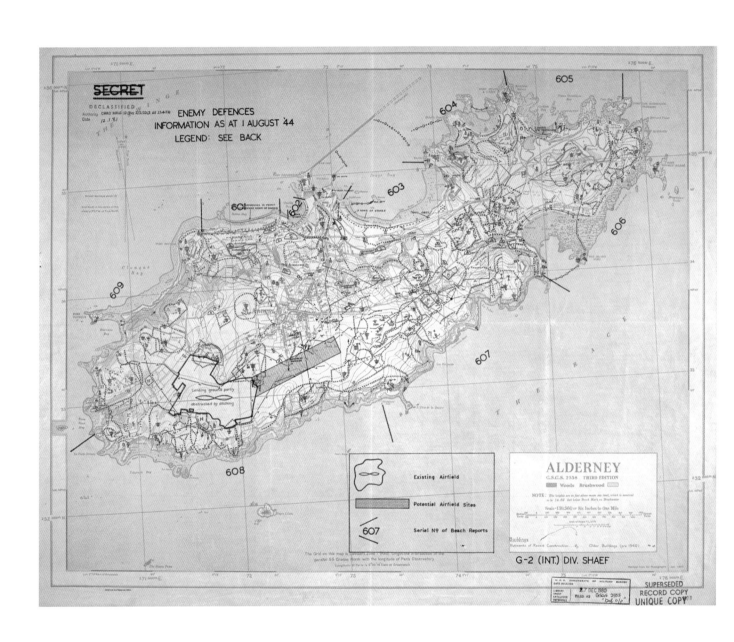

1944: A secret map of occupied Alderney

ALDERNEY IS THE northernmost of the large Channel Islands, which also include Guernsey and Jersey. For islands in some of the world's most heavily shipped waters, better mapped than most other places on earth, they occupy a curiously isolated position in British culture, far closer to France than to England, with a mixed heritage and naturally self-governing, insular attitude. They were also the only occupied part of the British Isles during the Second World War, seized by German troops in June 1940 following the evacuation of the British Army at Dunkirk on 3 June that year.

The symbolic significance of their occupation and the humiliation of Britain was capitalised upon by the German propaganda machine. Virtually the entire resident population was evacuated from Alderney prior to German landings, and from late 1941 the German army heavily fortified it for long-term occupation as part of Hitler's 'Atlantic Wall'. But what wasn't fully understood in Britain until 1945 was the system of concentration and labour camps built there, subsidiaries of the larger Neuengamme camp near Hamburg. It was occupied by prisoners including Jews and many from Eastern Europe.

One of the darkest chapters of the war to have happened upon British soil, just beyond the sight of Channel Islanders, the wartime events on Alderney have proven highly controversial. Allegations of mass murder were investigated after the end of the war, but there was virtually no remaining resident population to corroborate the claims. It was only in January 1981 that this 'secret' military sheet of Alderney was stamped as declassified. It is a 1943 revision of an older map, overprinted with details accurate up to 1 August 1944. Nothing about prisoners is mentioned, suggesting that up to this date exactly what was occurring on the island was unknown even to the British military.

The map shows massive fortifications: trenches, shelters, bunkers, obstacles, guns, minefields, radar stations and dumps leave a bewildering pattern across the island. These have been added to the map by cartographers working from detailed photographs taken during RAF sorties over the island during August 1944, but aerial surveillance can only show so much. Ditches, foliage and roofs may hide a great deal from the sky; bunkers can obscure as well as protect. All four of the camps later discovered – lagers Sylt, Borkum, Helgoland and Norderney (this latter is the large encircled area in the north-east, Sylt is directly below the red area) – are marked along with others as merely 'H' hutted camps.

The fact that Alderney was not a major target for recapture, or a point from which Germany could launch an assault upon the British mainland, may explain the tactic of mere close observation by British forces. Morale-boosting recapturing operations were contemplated, but never enacted. The fall of Alderney would not have altered the outcome of the war, but it would have diverted resources away from the Allied landings on 6 June (see D-Day, p. 108). Nevertheless, reference to beach reports and potential airfields (in red) on the map proves that plans were in place.

The island was blockaded by the Royal Navy, though reports of starvation were communicated even before. The German army evacuated in May 1945, but it was not until seven months later that the population was permitted to return.

1944: D-Day and the Battle for Caen

THIS IS A GSGS British military map of Caen, the historic Norman city that was once William the Conqueror's seat of power. Less than 10 miles from the coast, Caen was an Anglo-Canadian objective for D-Day. The symbolism was not lost on contemporaries: the largest seaborne invasion in history took the route of the last successful invasion of England – but in reverse. In the event, failure to punch through to Caen on 6 June allowed the Germans time to bring up reinforcements. The bloody two-month struggle that ensued cost thousands of lives – including those of French civilians. It also led to the destruction of much of Caen itself, bombed by the RAF (thus inadvertently providing cover for the German defenders) and later shelled by the Germans as they retreated. While Caen was in German hands it was impossible for the Allies to break out of their beach heads – it was difficult even to bring over sufficient reinforcements from the UK as there was simply no room for them; parts of Normandy resembled a giant military car park.

The competence of British and Canadian troops and their commanders – including the overall commander of Allied ground forces, General Montgomery – was called into question, and Montgomery's self-assurance tested Anglo-American relations to breaking point. Montgomery's caution was legendary, but the concentration of elite SS Panzer divisions he faced (upon a very narrow front) was seen nowhere else in the west. The Germans threw everything they had at Caen and fighting was ferocious. The Canadians and the Panzer Division Hitlerjugend (Hitler Youth) both stopped taking prisoners.[76]

This map was revised in March 1944 and printed in April, a matter of weeks prior to the invasion. Most military maps were printed in full colour, but this example is simply printed in black, red and blue, which made it more legible in the half-light of dawn. No military information at all is offered, despite months of intensive aerial reconnaissance (though the modern reader might be struck by the appearance of Beaulieu Prison, scene of a massacre of French Resistance fighters). Maps detailing the geology (critical in airfield construction, for example) and natural and military obstacles had been created by and for a tiny group of individuals with the top security clearance 'BIGOT' (a wry acronym, probably for British Invasion of German Occupied Territory). This version of the map offers as much help as a pre-war tourist map, and was carried by an officer with a vital role to play.

Later editions show the German defences in detail: attempts had been made to plot every observation post, pillbox, anti-aircraft gun, mortar, flamethrower, minefield and roadblock. A month's heavy fighting produced maps reminiscent of the trench maps of the Great War (see pp. 50 and 56).

The map is printed on thin paper – if Caen was captured on the first day it could be discarded soon enough – and it has been roughly folded. Carrying an item like this in Normandy was a danger in itself: German snipers seeking to pick off officers and NCO's became adept at spotting the gleam of sunlight off the perspex covers of map cases. One panel of this map is grubby, as though it has been stuffed in the map pocket of the owner's battledress. This particular example was found with items belonging to Lieutenant Colonel Jack Norris, who was responsible for the Royal Artillery element of the 6th Airborne on D-Day. The division was tasked with securing the British left flank. Landing by parachute and glider just to the north of Caen, the rivers Orne and Dives provided natural defences to the west and east, and the lightly armed paratroopers could concentrate their firepower to the south, facing an expected counter-attack from 21st Panzer Division. Norris's anti-tank guns were a critical element of the plan; the commanding officer of 6th Airborne, Major General Richard 'Windy' Gale, paid tribute to 'the architect of so much of my artillery defence with all its peculiarities due to our airborne limitations'. Norris landed with the first wave, but the all-important guns were due later in the day. Norris was watching them arrive when he was badly wounded. In Gale's words:

> The closing incident for this great day was the fly-in of the 6th Airlanding Brigade … hundreds of aircraft and gliders: the sky was filled with them … It is impossible to say with what relief we watched this reinforcement arrive. The German reaction was quick … Unfortunately at my headquarters poor Jack Norris … received a terrible throat wound; none of us thought that he could possibly survive, but he did. His loss to us out there was great.[77]

Norris's part in the landings was over but the plan was a success, thanks to the tenacity of the airborne troops.

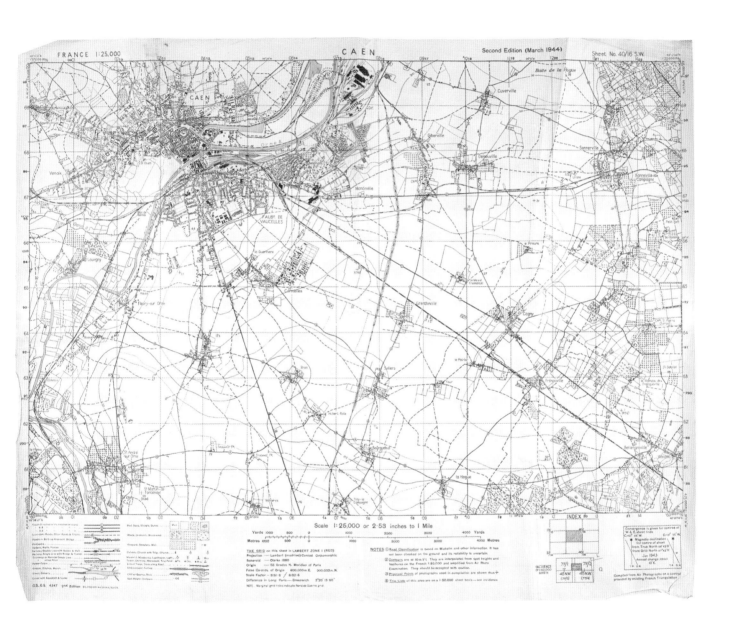

CHART IV

Occupation of Women by Regions, 1931

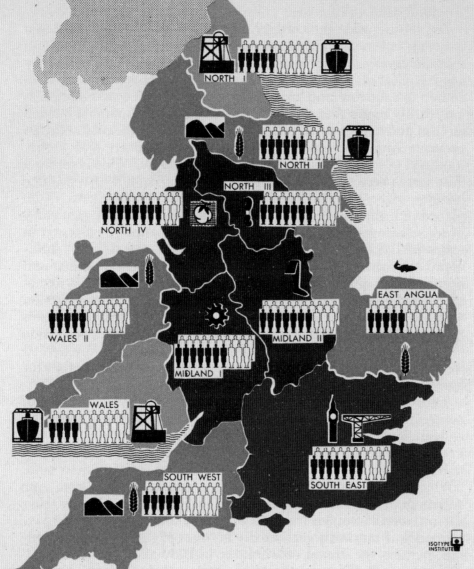

Each woman symbol represents 1 in 20 women in the coloured areas

black: occupied   outline: not occupied

NORTH III:	woollen textiles, etc.	NORTH I:	mining, ship-building, etc.
NORTH IV:	cotton textiles, engineering etc.	NORTH II:	agriculture, ship-building
MIDLAND I:	engineering etc.	EAST ANGLIA:	agriculture, fishing
WALES II:	agriculture, tourist centres	MIDLAND II:	hosiery etc.
WALES I:	mining, ship-building, etc.	SOUTH EAST:	government, docks,
SOUTH WEST:	agriculture, tourist centres.		commerce and all other trades

67

1945: Isotype maps and women in the workplace

THIS IS AN Isotype map from *Women and Work*, a 1945 publication in the *New Democracy* series by economist Gertrude Williams.[78] The maps by the Isotype Institute are integral to the book, given prominence on both title page and dust jacket. The *New Democracy* series reflected the mood of a nation that swept a socialist government into office in the landslide Labour victory of 1945. There was a perception – spelled out in the preface to this book – that many of the promises of 1914–18, including 'homes fit for heroes', had gone unfulfilled and that a second awful conflict had been the result: 'this time the democracies must not only win the war, they must win the peace'. The key to that, according to Williams, was active citizenship: full – and fully informed – participation in the decision-making process by every individual.

Isotype (International System Of TYpographic Picture Education) had been developed to meet just such a challenge. Established in 1920s Vienna by economist, social scientist and socialist Otto Neurath, and the artist Gerd Arntz, Istoype came to Britain via the Netherlands, as its creators sought to stay beyond the reach of Nazi persecution. 'Words divide, pictures unite' was one of Neurath's catchphrases. Isotype offers a visually memorable and accurate alternative to columns of statistics and dense exposition that are only comprehensible to the specialist reader. One of the basic tenets of Isotype design is that quantities are expressed through rows of identical symbols, rather than by enlarging or diminishing the size of a single symbol, making it easier to compare them with precision.

That is exactly what can be seen here. The figures are derived from the 1931 census, the most recent available as no census was taken in 1941, even though dramatic changes had taken place during the war. The caption explains: 'Each woman symbol represents 1 in 20 women in the coloured areas'. Symbols shaded black represent 'occupied'; those in outline, 'not occupied'. The colour coding represents the major industries of each region, including textiles, shipbuilding, agriculture, tourism and government. This is supported through use of simple symbols, such as a pithead (mining), a cog wheel (engineering), a stocking (hosiery) and Big Ben. The language of the map, for all its statistical basis, is nuanced: the shaded figures are positive, while the empty ones represent unfulfilled potential. The iconography of the symbols is timeless, remaining clear after seventy years.

The First World War had brought more women into the workplace than ever before, and in ways that would have been all but unimaginable a few years earlier (employed as police officers, for example). By 1939 the proportion of women in the workforce hadn't changed appreciably (roughly one third in London) and most women were once again employed in female-dominated sectors, some of which, such as nursing, were relatively traditional. However, there had been an enormous shift away from domestic service and into the factories, despite resentment that cheap female labour undercut men and kept all wages depressed. The state of the labour market in the 1930s favoured the employer. In the Second World War women once again assumed skilled roles that had been exclusively male in peacetime and in 1945 there was cautious encouragement for some of them to stay. A deep-rooted desire by government, demobbed servicemen and by many working women themselves to restore a sense of social stability as swiftly as possible by returning to pre-war norms was outweighed by a chronic labour shortage. One government solution was to invite citizens from Empire and Commonwealth countries to settle in the UK – with far-reaching consequences – but in terms of women in the workplace, the world of 1939 was gone for good. As the manufacturing industry declined, more women found office work, and in the 1940s major institutions began scrapping the marriage bars that had automatically ended many promising careers at the altar. The pay gap remained huge: the differential was narrower in professions such as teaching, but on average a woman earned half the wage of her male counterpart. However, things had begun to change. Gertrude Williams advised her readers to 'guard against too intense a concentration of activity within the narrow confines of the home', as employment profited both women and the community. Some of her readers were ready to listen.

1945: Occupied Berlin: the U-Bahn and daily life in the ruined city

THIS SEPTEMBER 1945 U-Bahn diagram is almost certainly the earliest post-war map of the Berlin underground railway network. It also shows the S-Bahn overground connections and the remnants of the rest of Berlin's transport infrastructure, the trams and buses. It is an unofficial map, not issued by the system operator, the Berliner Verkehrsbetriebe (BVG), but privately printed by obscure publisher M. & R. Meier. The printing costs were offset by the 50-pfennig cover price and by advertising – filling a gap in the market, perhaps, but no mean feat of entrepreneurship in the rubble of Berlin. The U-Bahn had ground to a halt altogether when the power generators failed in April 1945, days before the war's end, but trains were running again by June.

The map has been specially commissioned, rather than adapted from the standard BVG edition. It bears the monogram of Heinz Schunke, a draughtsman located in Offenburg in south-west Germany. The information is up-to-the-minute. Adolf-Hitler-Platz, for example, has already reverted to its original, pre-1933 designation Reichskanzlerplatz (preserving the name of the office, though not the individual). Unlike any known BVG map, including a proof date-stamped October 1945 which seems to have remained unpublished, Schunke's map indicates which sections remained closed.

The network had suffered from Allied bombing – the shallow cut-and-cover tunnels were vulnerable – but a great deal of devastation was caused by the deliberate flooding of tunnels in the final phase of the Battle of Berlin. Accounts of the last few days of the war are notoriously incomplete and contradictory. Time passed strangely in a twilight underground world of cellars, tunnels and bunkers – military and civilian – while overhead the city was shelled into submission street by street. Hitler may have personally ordered the flooding as early as 27 April.[79] Fearing that Soviet shock troops might use the tunnels to outflank the Reichskanzlei and Reichstag, sappers from the SS

Division Nordland (mostly foreign volunteers – by 1945 over half the Waffen-SS was non-German) were sent to blow out the roof of the S-Bahn tunnel running beneath the Landwehr Canal. The explosion probably took place in the small hours of 2 May, more than twenty-four hours after Hitler's suicide and on the day that the remnants of the Berlin garrison either surrendered or made last-ditch attempts to break out. Some 25 kilometres of interlinked S-Bahn and U-Bahn tunnels were flooded, claiming the lives of an unknown number of soldiers and civilians who had sought refuge there. U-Bahn carriages were even turned into makeshift hospital wards. There was time to evacuate and the number drowned was relatively low, but there is no denying the fear and panic caused by the slowly rising water; the damage took months to repair.

For the benefit of U-Bahn passengers, the map also shows the partition of their city and their country. A German nightmare of the First World War (see fig. 1, p. 8) was realised after the Second: at Potsdam in August 1945 the Allies divided Germany into zones of military occupation. There was no possibility of a new generation recycling the dangerous betrayal myths of 1918: this time defeat was total and unambiguous, backed by the presence of foreign soldiers. A generation of British National Servicemen would see out their time with the British Army of the Rhine, although the army of occupation switched rapidly to its Cold War role of first line of defence against the Soviet threat.

Berlin, within the Soviet Sector, received special treatment: it was divided among the Allies, becoming a microcosm of the country as a whole. The zones are not marked on this U-Bahn map – they never were – and trains from West Berlin rattled through closed 'ghost stations' in the East until 1989.

Maps are two-dimensional representations of three-dimensional space, which distort or highlight aspects of that space for a purpose; by their nature they impose a sense of order that can seem artificial, but seldom more so than here. More than half the housing of Berlin was uninhabitable and more than a million Berliners were homeless. There was a lack of food, power and clean water, and also the terrible absence of friends and family – killed, captured or wandering the Continent among millions of other displaced persons. In their place the city teemed with strangers, often hostile, mostly from the Red Army. Among the thousands of acres of bomb-blasted buildings, personal property and personal morality often went by the board; rape and other acts of violence became commonplace. Across the Continent, survival was often more important than celebrating victory or reflecting on defeat. The very existence of this map is quite profound in its efforts to return to a semblance of normality: against this backdrop of misery and lawlessness its creator actually managed to drum up advertising from the Astoria.

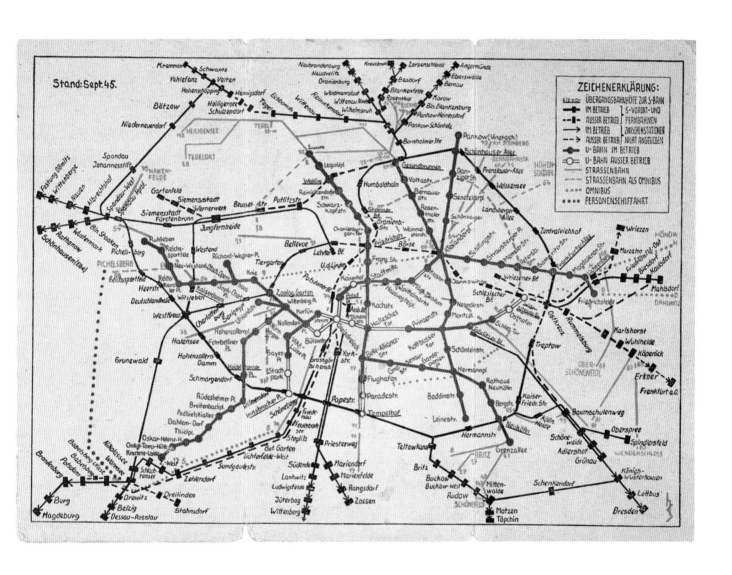

1945: Christmas greetings from Nuremberg

THIS US ARMY folded paper Christmas card was sent from the office of Robert H. Jackson, Chief US Prosecutor at the International Military Tribunal in Nuremberg. The city had great symbolic value for the Nazi regime (see Hitler's 'best *Parteitag*', p. 84) and its Palace of Justice was one of the few suitable buildings in Germany that remained largely intact. Jackson's personal assistant, Elsie Douglas, has written a brief seasonal greeting inside: 'Hello to you and a Merry Christmas!' A somewhat terse message, perhaps, but it was probably written at the end of November – a week after the commencement of the Trial of the Major War Criminals. British, American, Soviet and French judges and prosecutors tried Hitler's designated successor Admiral Dönitz and other leading Nazis, including Göring and von Ribbentrop, for crimes against peace, war crimes and crimes against humanity.

Jackson was noted for his 'ready wit' and the card is darkly humorous, possibly a release from the soul-sapping daily task of sifting through evidence of Nazi atrocities. Winged angels watch over the earth, joining together to illuminate the starry sky with a candle.

However, in this case they wear the white helmets, greatcoats and armbands of American Military Policemen, the 'snowdrops' that could be seen in the courtroom every day. There are echoes here of the United Nations globe on a light blue background, which may not be coincidental: although the UN flag was adopted in 1946, its designer, Donal McLaughlin, also headed the team that designed the Nuremberg courtroom. It is probably safer to say that this is cheap and cheerful single-colour printing, and that the MPs are a reference to the famous Nuremberg angels rather than America's new role as world policeman.

For the first time leading figures of an internationally recognised regime were held accountable for their actions. The trials were the official counterpoint to the private vengeance being enacted across the Continent. Indeed, they were a beacon of order and protocol in a savage post-war Europe which had – in regions that had been heavily fought over – reached a state of lawlessness unknown for centuries. As already seen in the case of Berlin (p. 112), in parts of Europe money gave way to barter, industrial and agricultural production were negligible, and personal property and morality counted for little in the face of lack of food, shelter and information about loved ones who might be dead, interned or displaced. To complete this picture of misery, a bitter winter was setting in.

Yet it is noteworthy that only the most senior figures (for the most part) were tried at all. It has been suggested that this allowed the majority of Germans to transfer responsibility for the events of 1933–45 to the heads of a few aberrant individuals,[80] from where it is a dangerously short step to presenting the majority of the population as victims of Nazism. Despite rhetoric about de-Nazification, it was rapidly apparent that denying all ex-Nazis office would bring the new Germany to a standstill. Soviet Russia was the greater threat, and there seems to have been little appetite in 1945 for pursuing every alleged criminal. The desire for normality was greater than the desire for swift vengeance.

This pattern would be repeated throughout Western Europe. In the heat of liberation summary justice was meted out to some perceived collaborators and war criminals. Far more common, however, was the spiteful shaving and humiliation of women accused of 'horizontal collaboration' and the ostracisation of their children. The example of the Channel Islands, the only part of the UK to have been occupied, seems typical: there were revenge attacks against 'jerry bags' but no islanders were prosecuted under the 1940 Treachery Act, and the SS guards from the Alderney concentration and forced labour camps were treated like any other POWs. The Nuremberg trials set an important precedent, but it would be twenty years before Europe had the stomach for a second wave of war crimes trials, this time spearheaded by private individuals such as the Nazi hunter Simon Wiesenthal.

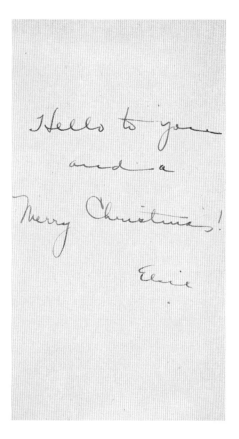

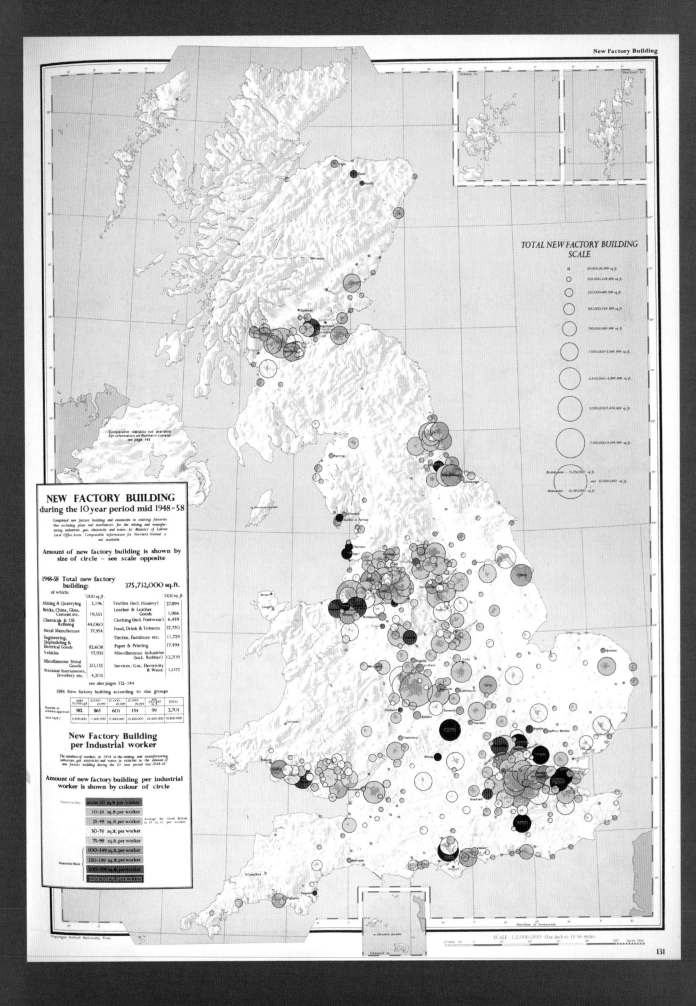

TOTAL NEW FACTORY BUILDING
SCALE

- 10,000-99,999 sq.ft.
- 100,000-249,999 sq.ft.
- 250,000-499,999 sq.ft.
- 500,000-749,999 sq.ft.
- 750,000-999,999 sq.ft.
- 1,000,000-2,499,999 sq.ft.
- 2,500,000-4,999,999 sq.ft.
- 5,000,000-7,499,999 sq.ft.
- 7,500,000-9,999,999 sq.ft.

Birmingham – 15,256,000 sq.ft.

over 10,000,000 sq.ft.

Manchester – 10,481,000 sq.ft.

Comparable statistics not available
for information on Northern Ireland
see page 145

NEW FACTORY BUILDING
during the 10 year period mid 1948-58

Completed new factory building and extensions to existing factories
(but excluding plant and machinery), for the mining and manufacturing industries, gas, electricity and water, by Ministry of Labour
Local Office Areas. Comparable information for Northern Ireland is
not available

**Amount of new factory building is shown by
size of circle – see scale opposite**

1948-58 Total new factory
building: 375,732,000 sq.ft.
of which:

	'000 sq.ft.		'000 sq.ft.
Mining & Quarrying	2,196	Textiles (incl. Hosiery)	27,894
Bricks, China, Glass, Cement etc.	19,553	Leather & Leather Goods	1,966
Chemicals & Oil Refining	44,060	Clothing (incl. Footwear)	6,459
Metal Manufacture	37,954	Food, Drink & Tobacco	37,770
Engineering, Shipbuilding & Electrical Goods	82,608	Timber, Furniture etc.	11,729
		Paper & Printing	17,494
Vehicles	47,933	Miscellaneous industries (incl. Rubber)	12,709
Miscellaneous Metal Goods	20,132	Services: Gas, Electricity & Water	1,072
Precision Instruments, Jewellery etc.	4,203		

see also pages 132-144

1954 New factory building according to size groups

	under 10,000 sq.ft.	10,000-19,999	20,000-49,999	50,000-99,999	over 100,000 sq.ft.	TOTAL
Number of schemes approved	982	865	601	154	99	2,701
Area (sq.ft.)	6,900,000	11,600,000	17,500,000	10,200,000	24,600,000	70,800,000

New Factory Building
per Industrial worker

The numbers of workers in 1954 in the mining and manufacturing
industries, gas, electricity and water, in relation to the amount of
new factory building during the 10 year period mid 1948-58

**Amount of new factory building per industrial
worker is shown by colour of circle**

Numbered blue	under 10 sq.ft. per worker	
	10-24 sq.ft. per worker	
	25-49 sq.ft. per worker	Average for Great Britain is 57 sq.ft. per worker
	50-74 sq.ft. per worker	
	75-99 sq.ft. per worker	
	100-149 sq.ft. per worker	
Numbered in black	150-199 sq.ft. per worker	
	200-299 sq.ft. per worker	
	300 and over sq.ft. per worker	

Copyright: Oxford University Press

SCALE 1:2,000,000 One inch to 31.56 miles.

Bust to Boom to Bust. 1946–1972

The Atlas of Britain.
Oxford: Clarendon Press, 1963

The *Atlas of Britain* was conceived as
a mid-twentieth-century snapshot during
what was believed to be a period of rapid
social and economic change. Its large,
precise colour maps visualised statistics on
industrial productivity, demographics and
agriculture. Produced by a conglomerate
of academics and government departments,
the atlas constituted a proud symbol of
Britain's 'golden age.'

From the end of wartime hostilities came new and often portentous beginnings. International consensus arrived in the form of the United Nations (1945), while the boundaries of the new state of Israel and a thick pencil line creating Pakistan out of India (1948, 1947) were both defined on the map (see p. 118). Post-war reconstruction employed maps as well as surplus labour, for which, for expediency, the Ordnance Survey in Britain released compilation-only aerial photographs in 1949. It withdrew them for security reasons in 1951.

The ideological struggle between capitalist West and communist East – the Cold War – saw its first confrontation in Berlin with the Russian blockade of 1948. Further flashpoints included the Korean War (1950–3), the war in Vietnam (1959–75), the building of the Berlin Wall (1961) and the nuclear brinkmanship of the 1962 Cuban missile crisis. The Cold War hung like a cloud over the entire post-war period, acting as a reality check for the decades known as the 'golden age', which ended with the Middle Eastern oil crisis and economic crash of 1973–5 (see p. 163).

The 1950s are often presented as an era of prosperity in Britain, ushered in by the Festival in 1951. The National Health Service (NHS) was founded in 1948 and there was an end to wartime rationing in 1954. Queen Elizabeth II was crowned in 1953, the same year as the New Zealander Edmund Hillary with the Nepalese Tensing Norgay scaled Everest, the world's tallest mountain. Prosperity and rising confidence led to an era of redefined social values, a cultural revolution and increased personal freedom throughout the 1960s. This found expression in art and music (Elvis and The Beatles), in civil rights (particularly in the US), and in environmental, feminist, anti-war and anti-nuclear protests. Capital punishment was abolished in Britain in 1969.

The legacy of war infused the post-war period. The science behind the Manhattan Project informed nuclear weapons testing throughout the 1950s until the Test Ban Treaty of 1963, while nuclear power was developed as an alternative energy source to oil up to and even beyond the Chernobyl nuclear disaster in 1986. Technologies developed for Nazi German missiles went on to inform the United States space programme inaugurated by President John F. Kennedy in 1961, which succeeded in placing a man on the moon in 1969 (see p. 154). British wartime code breakers at Bletchley Park provided the basis for modern computing.

During this period Britain faced up to its post-colonial status. India, its jewel, declared independence in 1947. Attempts to seize the Suez Canal Zone from Egypt in 1956 led to a humiliating climbdown. Its foreign mapping agency, the Department of Colonial Survey, substituted the word 'Colonial' with 'Overseas' in 1957. Britain redefined its empire as a Commonwealth of Nations in 1949. Immigration from its former colonies started in earnest, and Britain started to become more multicultural. Politically it strove to align itself with the United States, while at the same time attempting to strike a balance with Europe. Its entry to the European Economic Union, however, was barred in 1968 (see p. 146).

The 'white heat of revolution' was British prime minister Harold Wilson's 1964 description of Britain's technological transformation. The 1960s signified science, industry and modernisation, and this is captured in the 1963 *Atlas of Britain*, with its high-quality colour photolithographic maps containing economic data. Technical colleges sprang up, and scientific geography began to be taught in universities. Maps benefited from increasingly sophisticated production methods, while mapping was practised and theorised as a purely scientific act, a reaction perhaps to the highly subjective propaganda cartography of the war years. Mapping as used in advertising was barred entry to the cartographic canon.

Technology played its part in the process called globalisation, whereby lines of communication, travel, media and economics became more homogeneous the world over. This meant that when the 1973 Middle Eastern crisis threatened the supply of oil (see p. 163), the debilitating effects sent shudders through the developed world.

1947: Partition and the end of empire in India

THIS IS AN unofficial souvenir map that commemorates one of the bloodiest and most divisive episodes in British or Indian history: the partition of the subcontinent which accompanied the British withdrawal. Our map is confined to the province of Bengal; it was published locally by the Dipti Printing and Binding Works, Calcutta, and advertises their services.

The boundary itself is Cyril Radcliffe's 'shadow line', the official border that was confirmed only on the day following independence – a largely unsuccessful attempt to prevent bloodshed and disorder. Radcliffe was a high-flying British lawyer, whose lack of prior experience of India (or of drawing international boundaries) was initially regarded by all parties as an asset, as he was considered impartial. Radcliffe was distraught at the consequences of his decisions and refused to collect his fee. Even the thickness of the pencil he used was called into question as (given the scale of the maps he used) whole villages became disputed territory.

The map itself has a positive tone. The crossed flags of Pakistan and India suggest a neighbourly accommodation, and the statistical analysis itemising the density of the Muslim and non-Muslim population per square mile appears to justify the logic of partition. It was a partition that had been tried and abandoned in 1905, within the context of British India, and that this time would create an international border between the regional capital Calcutta, a centre of the jute industry, and its jute-growing hinterland. Taken across the subcontinent as a whole, partition disrupted an administrative structure and infrastructure that had developed over centuries, and in economic terms it benefited no one. However, this map is evidence that by mid-1947, at a popular level, what had once been described by Gandhi as the 'vivisection' of India had come to be regarded as something to be commemorated – to be pinned to the wall.

Partition had certainly come to be regarded by all parties as the lesser evil, and possibly the only viable alternative, to civil war. It was clear that Britain's Labour government had neither the ideological will nor the financial means to delay independence further and, rather belatedly, had come to believe that an autonomous Dominion of India could be a useful ally in Asia. The main obstacle to a smooth transition of power was the communal tensions between Hindus, Muslims and Sikhs, which had so often been played out to the advantage of the Raj in the past.

Mohammad Ali Jinnah's Muslim League, which had manoeuvred itself into a position whereby it could claim to speak for the majority of Indian Muslims, demanded regional autonomy for Muslim-majority provinces. Jinnah spoke dismissively of a 'moth-eaten Pakistan', which would leave almost as many Muslims in India as in the new state. Jawaharlal Nehru, leader of the increasingly Hindu-dominated Indian National Congress, also wanted to preserve India's territorial integrity, but not at the expense of devolution, which would challenge his vision of a centralised, modern, socialist state. Against a background of strikes, mutinies and communal riots, the charismatic new Viceroy, Earl Mountbatten, decided that speed was the best way to avoid a complete breakdown in law and order.

The British had been in India for more than three hundred years. In setting independence for midnight on 14–15 August 1947, Mountbatten gave them precisely seventy-three days to leave. All parties were aware of the likelihood of further bloodshed, but no one envisaged the scale of the bloodbath that ensued. Over 11 million people fled their homes, in one of the largest migrations in human history, and the number of dead is generally estimated at between 500,000 and 1 million. The worst of the violence took place in Punjab which, like Bengal, was divided in two. However, in the aftermath of the Great Calcutta Killings of August 1946 – several days of sectarian slaughter, an initiation for the subcontinent into the horrors to come over the next twelve months – Gandhi had taken up residence in the city. Although he was becoming increasingly sidelined by mainstream politicians his presence in Calcutta during the transition had a calming effect locally. The positive tone of this particular map becomes more explicable in the knowledge that Bengal was spared some of the horrors that afflicted other parts of the country.

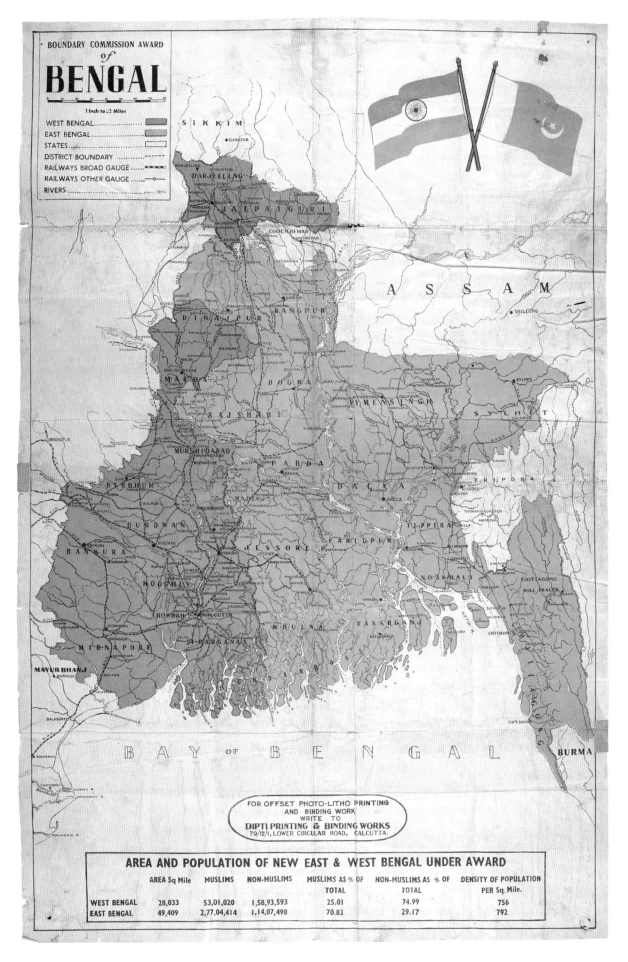

1949: Two maps charting the changing face of British railways

THE WEALTH AND confidence of the 'big four' rail companies was apparent in advertising posters of the interwar years. Containing some of the most iconic views of the British landscape – Snowdonia, Durham Cathedral and Brighton Pier – these large, eye-catching images influenced the itinerary of the British holidaymaker. But the picture had altered by 1945. The war effort had run the network into the ground, nationalisation in 1948 becoming one answer to a chronic lack of investment. These were tough years for all. However, austerity is given a positive twist in Estra Clark's 1949 railway poster map of York-shire. Lithographed in a traditional painterly style, the map attracts by emphasising the traditions and values of Yorkshire through heraldry,

historical scenes and the family unit. It appeals to a localism by calling upon the most traditional geographic entity, the English county.

By the 1960s such insular, antiquated localism was threatened by modernisation. People's mobility had increased hugely and their holi-days were changing. The first section of the first motorway in Britain, the M6, was completed in 1958, by which time cheaper air travel was also more available. Over the course of the 1960s, Britain's sprawl-ing railway network was reduced by far more than the 2,000 stations that Dr Richard Beeching's *Reshaping of the British Railways* had recommended. To the modernisers, the dead wood was precisely those quiet, rural and, at least to the new business (as opposed to service) model, unprofitable stations.

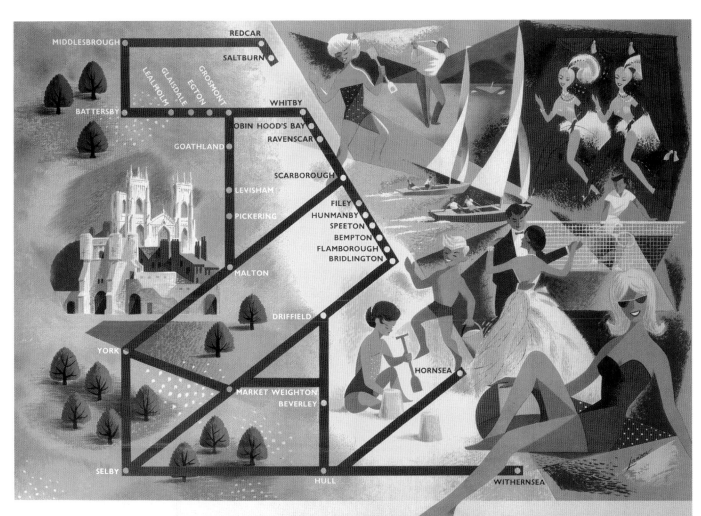

EXPLORE THE YORKSHIRE COAST BY TRAIN
AND NEARBY COUNTRYSIDE

Something of a need to demonstrate otherwise is present in the poster advertising the Yorkshire coast produced for British Railways North Eastern Region by the graphic designer R. M. Lander in 1963, the same year as Beeching's report. Contrary to depicting a sleepy back-water, the poster presents a thriving Côte d'Azur, with the York-shire coast railway its communicating artery. Bright colour tints, sharp lines and sun-drenched beaches, connected by an incisively rendered rail network à la Harry Beck, provides a modern edgy impression, demonstrating the line's commercial acumen and, with it, its right to survive.

The truth was that by 1963, local excursions to Scarborough or Bridlington, York or the North York Moors could be made by car or bus, door to door, far more conveniently with heavy luggage. Although a station opened in 1947 to service the newly opened Butlins holiday camp at Filey, the first closures on the Yorkshire coast line began soon after. The network was zealously pruned, a number of coastal stations closing in 1970 and large sections from Whitby in 1965. Yet eight years later, part of the inland line from Grosmont reopened as a heritage steam railway. The quaint backwardness that had been anathema to the 1960s mod-ernisers had quickly become a popular and lucrative attraction in itself.

1949: Aquila Airways:
austerity, government surplus, and the post-war travel boom

FEW MAPS ENCAPSULATE the spirit of an era quite so comprehensively as this one. In every respect it represents determined attempts to return to normality, if not exactly to pre-war levels of opulence, while actually making do with very little. As a physical object, it is a triumph of simple three-colour printing, the essence of cheap and cheerful. The map-maker, pioneering aviator and round-the-world yachtsman Francis Chichester, had established his company after a wartime stint as a navigation expert by bulk-buying 15,000 surplus Air Ministry maps and pasting them onto board to make jigsaws.

Aquila Airways was a tiny but remarkable outfit, recently described by one commentator as a 'seat-of-the-pants swashbuckling enterprise'.[81] Operating converted war-surplus Sunderland flying boats that were available relatively cheaply – in 1945 numbers of newly built aircraft were simply taken out to sea and scuttled as no other use could be found for them – Aquila was established by former Coastal Command pilot Wing Commander Barry Aikman, who had flown the type during the war.[82]

The fledging airline cut its teeth under government contract during the Berlin Airlift of 1948, the only known operational use of flying boats within Central Europe. There was a long tradition of under-equipped British forces looking to the private sector – from the Coca Cola lorries and Pickford's removal vans hired to transport munitions during the Suez Crisis (see p. 134) to the leasing of the liner *SS Canberra* as a troop ship during the Falklands War. Stalin's blockade enabled Aikman to expand his fleet from two aircraft to a dozen, and he swiftly moved from freight to passenger transport.

Aikman had identified a gap in the market perfectly tailored to his expertise and experience, and which did not (and could not) clash with British European Airways routes. His flying boats could take off from the traditional site of Southampton Water for exotic destinations such as Madeira and the Canaries, where a lack of concrete runways precluded the use of BEA's conventional aircraft. 'Eagle wings to warmth' was the slogan cited in one brochure from the 1950s,[83] which also described the eight-hour flight to Funchal in Madeira in a converted warplane as 'one of the most glamorous experiences in European air' – precisely what Aikman was aiming for.

In some ways Aikman was remarkably prescient: Aquila was the first airline to make serious efforts to open up a route to Ibiza, although the project was blocked by the Spanish authorities. However, flying by Aquila was only for the affluent: a ticket price of £80 equates to a four-figure sum today. Aikman was attempting to recapture something of the glory days of Imperial Airways as he had known it in his youth, but there was also a broader appetite in ration-weary Britain for foreign travel that was just waiting to be tapped.

In 1950, a year after Aquila's first passenger flight, Russian-born entrepreneur Vladimir Raitz chartered a Dakota and flew about twenty people to Corsica to stay under canvas (mostly left behind by the US personnel who constructed Calvi airport in 1943) and eat two daily meals that contained meat; the package holiday was born. In the face of all hopes and expectations, victory in Europe had ushered in a period of greater shortages than ever before, as basic foodstuffs such as bread and potatoes joined the list of rationed items for the first time. Raitz's Corsican meat-meals were a real draw in their own right.[84]

However, travel at this time was still beset by virtually impenetrable levels of bureaucracy, chiefly created by the Ministry of Civil Aviation and mostly designed to protect the nationalised carriers BEA and BOAC – although factors such as strictly controlled personal foreign currency allowances also played a part. Much of it would seem ludicrous to subsequent generations. For example, the Ministry granted Raitz permission for his Corsican venture on the proviso that his holidays were exclusively for teachers and nurses. It would have been difficult for anyone in 1950 to predict the mushrooming of mass tourism to new destinations such as Benidorm on the Costa Blanca (a marketing conceit that became a geographical term, devised by someone at BEA) by the end of the decade.

As for Aquila, the airline was purchased by larger independent operator Britavia in 1954, but the days of the big commercial flying boats were numbered; operations ceased in 1958, less than a decade after they had begun. Nevertheless, our map represents a landmark moment in post-war leisure and travel.

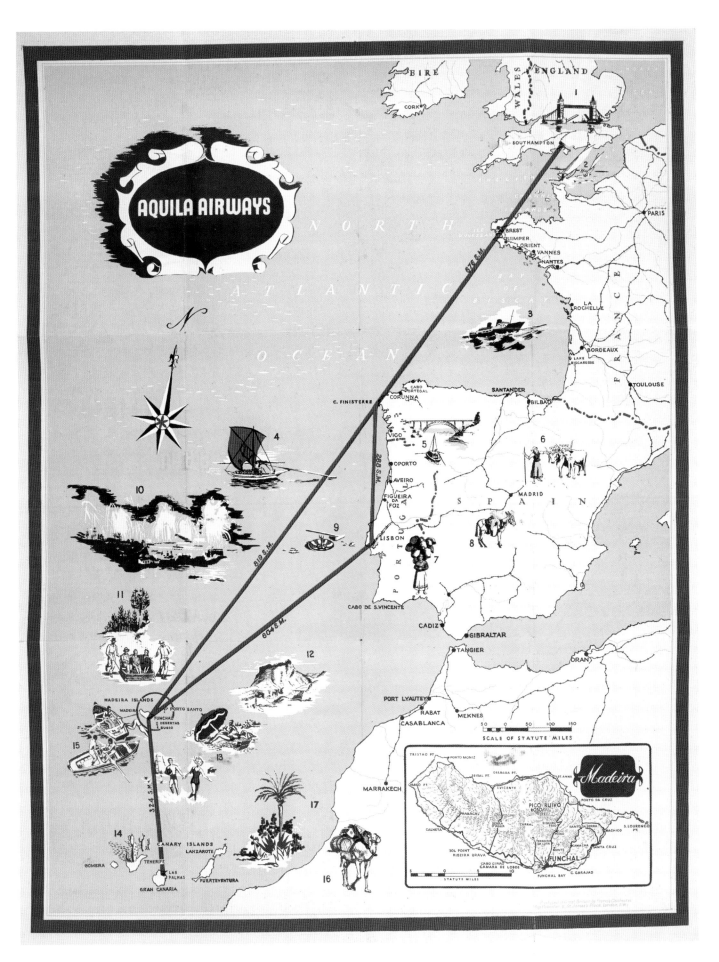

1951: What do they talk about? The British and the Festival of Britain

THE FESTIVAL OF Britain was planned to mark the centenary of the Great Exhibition, the early world's fair held in the Crystal Palace which symbolised high-Victorian confidence and industrial progress. According to its organiser, dubbed 'Lord Festival' (Labour minister Herbert Morrison), it was to be a 'pat on the back' for the British people after a thoroughly miserable decade. The queues, shortages, power cuts and rationing had worsened post-war, and the festival was a modest opportunity to look to the future through a celebration of British architecture and design. Britain would be remade. The main festival site was 27 acres of bomb-blasted warehousing on the South Bank in London, but it was to be a truly national occasion, with exhibitions in Glasgow and Belfast, and touring musical and dramatic performances.

It is that sense of national involvement which the artist Cecil Walter Bacon conveys through this poster. 'What do they talk about?' is Bacon's wry but optimistic take on the state of the nation in 1951. Bacon himself was a successful commercial artist. A longstanding contributor to the *Radio Times*, he also worked for London Transport, British Railways and the Post Office. Given away as a souvenir with the festival issue of *Geographical Magazine* (the magazine of the Royal Geographical Society), additional copies of this map could be ordered at 2 shillings and 6 pence. It is worth noting that the major sponsor was Esso. Petrol was still rationed – indeed it was still pooled (branded petrol was not sold again until 1953) – but the major players were already seeking to revive brand loyalty. Unusually for a petrol map, Bacon doesn't show a single road or name a single town. We are invited to guess through what the locals are talking about.

London, of course, is dominated by the festival site, the Skylon and Dome of Discovery, and the Royal Festival Hall, one of Britain's most exciting post-war buildings to date, and a symbol of renewal. Bacon's choice of subject matter is frequently topical. He features Esso's Fawley Refinery, rebuilt and expanded in 1951 and still the largest in the UK. Oxford scholars can be seen discussing the old (Aristotle) and the new (Lord Nuffield): endowed by the founder of Oxford-based Morris Motors, the first bricks of Nuffield College were laid in 1949. He also celebrates the ill-fated Bristol Brabazon. A giant passenger airliner conceived for transatlantic flights, the prototype had been test flown in 1949 but was broken up for scrap in 1953 due to lack of commercial interest (although the technology was applied to other projects). Bacon's holidaymakers wear the daring new two-piece bathing costumes (the bikini proper had arrived in 1946). Not everything new is good, however: the lament in Scotland is 'Whisky for Dollars'. The priority for successive post-war British governments was to bring in foreign currency in order to reduce the balance of payments deficit. Industry was no longer engaged in war work, but very few manufactured goods found their way into British shops: anything which could be sold abroad was exported, and Scotch was no exception.

The marching Orangemen of the Northern Irish unionist Orange Order may strike the modern viewer as out of place in a gently comic map of this nature, but there's no reason to think that Bacon was being provocative. Captured German records had revealed the full extent of IRA involvement with the Abwehr, and de Valera's official message of condolence on behalf of the Irish government on the death of Hitler had also been made public. Support for the nationalists, even north of the border, had reached an all-time low; the Orange Order dominated Northern Irish politics, and the Troubles were twenty years in the future (see p. 160).

Other topics featured on the map, including repeated references to the weather, are either typically but uncontroversially British, or record a Britain which was already vanishing. The title of the map – what do they talk about in factories, mills and offices – reflects this transformation. Industrial Britain was in decline; 'office world' was on the rise. In festival year, Britain was on the cusp of something new.

1952: A map of Caribbean oil

WITH THE WORLD'S, especially the United States', increasing consumption of oil after 1945 came a greater dependency upon the oil-producing regions of South America. Although the mass exploitation of Middle Eastern oil began in earnest during this time, American companies already enjoyed favourable concessions, a reliable stream of oil, and influence over Latin American national governments and their economies.

But as the Cold War gathered pace the US faced an increasing battle against nationalist and communist sentiments in Latin America. It had to protect its oil supply; in doing so, it had financed and supported sympathetic, corrupt governments, and wasn't above supporting the overthrow of unsympathetic ones, as it did in Guatemala in 1954. Such actions would harden opinion against it, as witnessed in Caracas in 1958 (when the US vice-president Nixon's car was pelted with stones) and in Castro's Cuban revolution one year later (see Bahía de Cochinos, p. 144).

A Texas trade publication called the *Oil Forum*, run by the enigmatic Brian Orchard Lisle, an English-born Second World War pilot and later kayaking champion, published journals, books and extremely large colourful wall maps for the vast and expanding oil business to visualise and assist strategy, planning and awareness. Maps of Middle and Far Eastern oil regions would follow over the 1960s, but Lisle's earliest map of Caribbean oil from 1952 shows the all-important area of American strategic, political and economic interest in bright and lucrative reality.

Although entitled 'Caribbean oil', the map concentrates on the north-western area of South America, with Venezuela, Colombia and Peru dominant, and islands such as Britain's own oil-producing Trinidad shown as insets. Geological colouring indicates favourable, unfavourable and improbable mineral deposits: in the key this is called a 'petroliferosity index'. An accompanying 'Guide to politico-economic climate' provides prospectors with the relative favourability of national governments. Oilfields, refineries and pipelines are marked in red; these are clustered mainly around the Lake Maracaibo and Barcelona regions of Venezuela, where oil had first been discovered in 1914 by the British–Dutch-owned Royal Dutch Shell. Vast untouched portions of the Colombian interior are shown as ripe for exploitation.

Wars and cars, and the worries of overpopulation (expressed in the 1958 land use map, p. 142), lent added impetus to overcoming natural and political obstacles to oil. The latter proved as problematic to the United States as the Amazon rainforest. In 1943 the new Venezuelan democratic government's petroleum law altered the previously unfavourable cut of oil profits that had kept the majority of the country in poverty. Yet the new fifty–fifty cut in profits remained a far better option for Creole, Shell, Texaco, Standard and Gulf Oil than full nationalisation. Money was still to be made and security to be ensured against the threat of creeping communism.

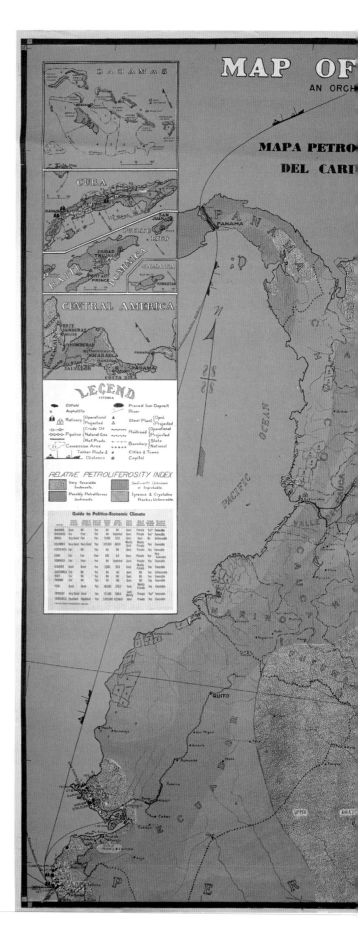

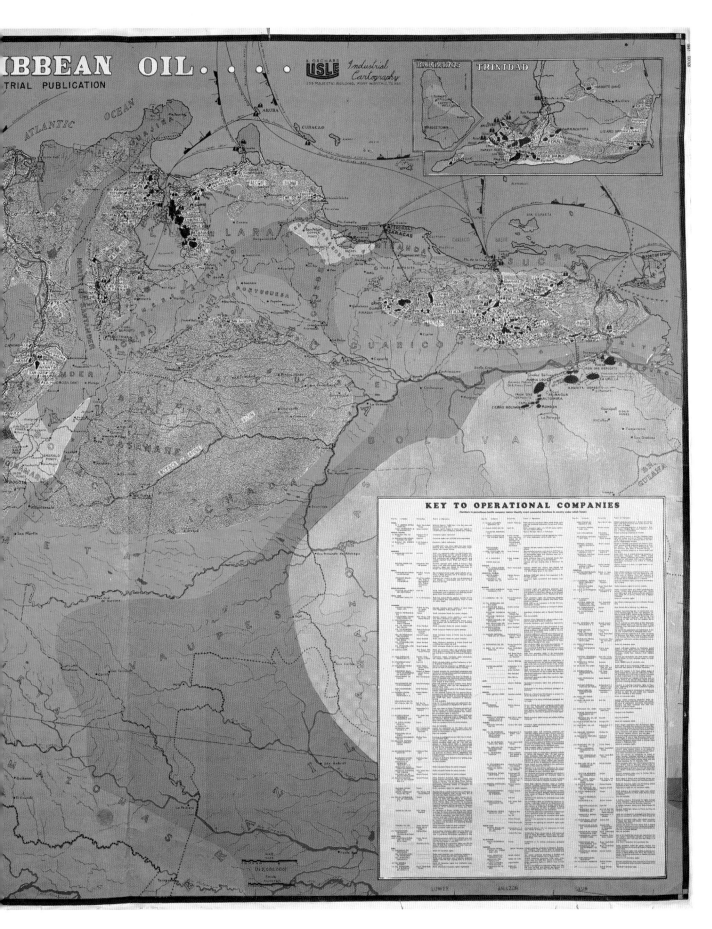

1953: The royal tour of the Commonwealth: an itinerary for the admiral

T HE 1953–4 ROYAL tour of the Commonwealth was an opportunistic and highly successful publicity exercise. It introduced Elizabeth II, the newly crowned Queen and head of the Commonwealth of Nations, to this collection of newly independent former British colonies. The exercise salvaged both an earlier tour of New Zealand cancelled due to King George VI's ill health and the 1952 replacement tour by Princess Elizabeth, also cancelled following his death, and turned it into a key promotional event. The image of the Queen as a benevolent figurehead of similarly young nations was a powerful device in positioning Britain and its monarchy in a post-imperial setting.

It was also the inaugural 'modern' British royal tour in the sense that it was the first in the age of modern media, complete with choreographed royal walkabouts broadcast across the world on Pathé newsreels. A BBC transmitter was placed on board the specially refitted Shaw Savill liner SS *Gothic*, and there was extensive coverage in newspapers.[85] Logistically the tour was a complex operation carried out under the public gaze: a six-month, 40,000-mile circumnavigation of the globe, beginning in November 1953 and taking in thirteen former colonies by a combination of land, air and (the bulk of the travel) sea.

For the Royal Navy, responsible for the safe passage of the Queen, the tour constituted a military operation, yet it retained its ceremonial edge. A special map of the tour was produced by the Hydrographic Office and issued in November 1953 to commanding officers of the Queen's vessels, along with typewritten notes on personnel and the

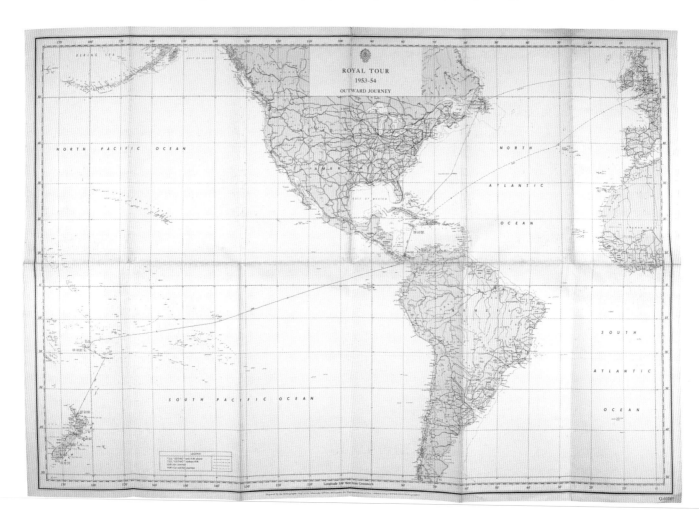

itinerary. It showed the projected track of the tour, together with dates and intersections, across the Atlantic to Bermuda, through the Panama Canal and on to Tonga, scheduled to reach New Zealand in time for Christmas. This copy belonged to Vice Admiral Conolly Abel Smith, commanding officer of the tour and captain of the Royal Yacht *Britannia* on her maiden voyage, the final leg of the tour through the Mediterranean Sea to Gibraltar and back to England.

Abel Smith was a highly decorated officer and his appointment was in recognition of his achievements. Presented to the admiral, this map provided the strategic (and confidential) overview of the tour, but also acted as a memento; its condition suggests that it was never used. Ceremonial performance informed much of this plan, especially in Australia and New Zealand which remained the real focus of the

tour. The navies of both nations had the honour of escorting the Queen in their waters.

The tour stayed for one month in New Zealand and two months in Australia, by far the longest stop-offs. After this, however, the route became rather more selective. From Australia the Queen visited Ceylon, omitting India which had declared independence outright in 1949. From there she travelled to Aden and south across land to Lake Victoria, giving the Suez Canal Zone – then occupied by British troops, as well as Cyprus – particularly wide berths. Visiting a place where Britain's influence was less welcome would have been counter-productive to the aims of the tour. The route of the return leg made unequivocal the post-colonial state of the world.

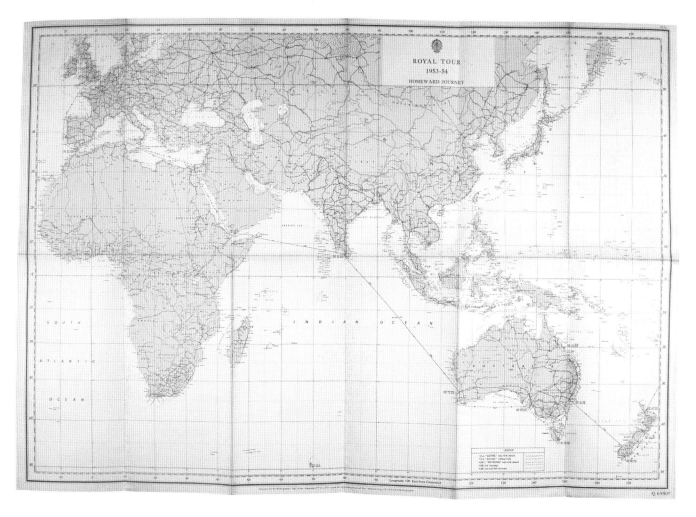

1953: Cruising on the 'Green Goddess': a transatlantic coronation celebration

THIS STRIKING POSTER advertised the Coronation Cruise on board the 'Green Goddess', the green-liveried Cunard passenger liner RMS *Caronia*.

Launched by the present Queen, then Princess Elizabeth, the *Caronia* made her maiden transatlantic voyage in 1949. She was a state-of-the-art vessel, fitted out with en-suite bathrooms and an open-air swimming pool. The golden age of liner travel (as opposed to cruising) was, however, all but over, thanks to competition from a new generation of long-haul jet airliners. After a decade in service the *Caronia* was refitted as a cruise ship, and within a quarter century of her launch she was broken up for scrap.

Only the most prescient passengers might have guessed at her fate in 1953. The Coronation Cruise was aimed at the wealthy American market. After a luxurious European cruise, detailed on the map, the ship docked in Southampton where she became a floating hotel for the duration of the coronation. On the day itself, her 500 passengers were conveyed by specially chartered Pullman train to London, where seats had been reserved for them at the purpose-built viewing stand at Apsley House, the former home of the Duke of Wellington.

The Anglo-American flags make perfect sense in context. Imagery of royalty also abounds, and London landmarks are gathered on the horizon. The baroque dolphins, flying fish and scallop shell all seem very 'new Elizabethan', and the depiction of the Spanish Armada clinches the reference, harking back to the perceived glories of the first Elizabeth's reign – some swashbuckling fun after an era of austerity. The artist has signed with his initials 'D. E. B.', a message on a bottle floating in the lower right foreground.

1954: Tolkien's map of Middle Earth: the entire story at a single glance

ONE OF THE key components of J. R. R. Tolkien's three-volume fantasy novel *The Lord of the Rings* is the map of the imaginary world in which the story is set. This general map of Middle Earth, over which the 1,800-page epic narrative roams, is folded into the back of the first volume. Now well known the world over, with 150 million copies sold, the book and its map have gone on to inspire a whole genre of fantasy writing, imagery and gaming culture. Though literary and imaginary in content, the map retained vestiges of the 'real' world in which it was created. And more than simply supporting the reading experience, it became in many ways a visual representation of *The Lord of the Rings* and the fantasy genre in general, as instantly recognisable a prototype as the British Empire or Underground map.

The most obvious reason for the map being so intrinsic to the book is that the tale is essentially one long journey (by the character Frodo Baggins) from one side of Middle Earth to the other. We use maps to follow journeys and Tolkien, a First World War veteran who had learnt to draw and navigate unfamiliar places through maps, fully appreciated the value of cartography as an aid to understanding place. He used it masterfully to embellish his own fantasy world.

Published in 4,500 copies by George Allen & Unwin in the first edition, first impression of *The Fellowship of the Ring*,[86] the map is an elegant offset lithographic print with red lettering, produced from a design drawn by Tolkien's son Christopher, which was created in turn from numerous sketches made by his father. It is a bird's-eye view, a suitably antiquated style of map, which also serves to accentuate the dramatically high mountains and other improbable geographical features. Tolkien was certainly no geographer, however. He was an Anglo-Saxon scholar and Oxford professor, who from 1917 had been working on a pre-history of the world, supposedly to compensate for the lack of a half-decent one in the English tradition.

Tolkien cleverly used the map to reinforce the reality of Middle Earth. He had it drawn in such a stylised way that it appeared actually to have originated there. The whole work, in fact, archaic writing style included, was constructed as a true-to-life transcription of events by the main characters.[87] The geography of Middle Earth was broadly similar enough to Europe – in coastal shape anyway – to support the pre-historical illusion. Many places shown on the map were never to be encountered in the story, suggesting that Middle Earth was real.

The map could exist independently of the book, and was redrawn and reproduced in different mediums to satisfy the increasingly vast commercial popularity of *The Lord of the Rings* in print, film, music and popular culture. Tolkien's own anti-urban, pro-rural ideology, which shone through in the ecological sub-theme of the book (talking tree-shepherds attacking industrialist wizards, to name but one episode), made his work extremely popular in the 1960s. The book was also able to capitalise on the limitations of the map: significant portions of the tale occur in forests and underground; and in the episode of greatest spatial confusion – the tortuous journey through dwarf mines – the lack of map for either party or the reader adds immeasurably to the sense of disorientation.

It is impossible to say how many times the average reader might have flicked to the back of the book to cross-reference a name or ponder the travel dilemmas of Frodo and friends (who incidentally never used maps). The map became endowed with the tales themselves. As the *Tolkien Encyclopedia* explains, 'while the story unfolds over hours of reading, the map allows the story to be recalled at a glance'.[88]

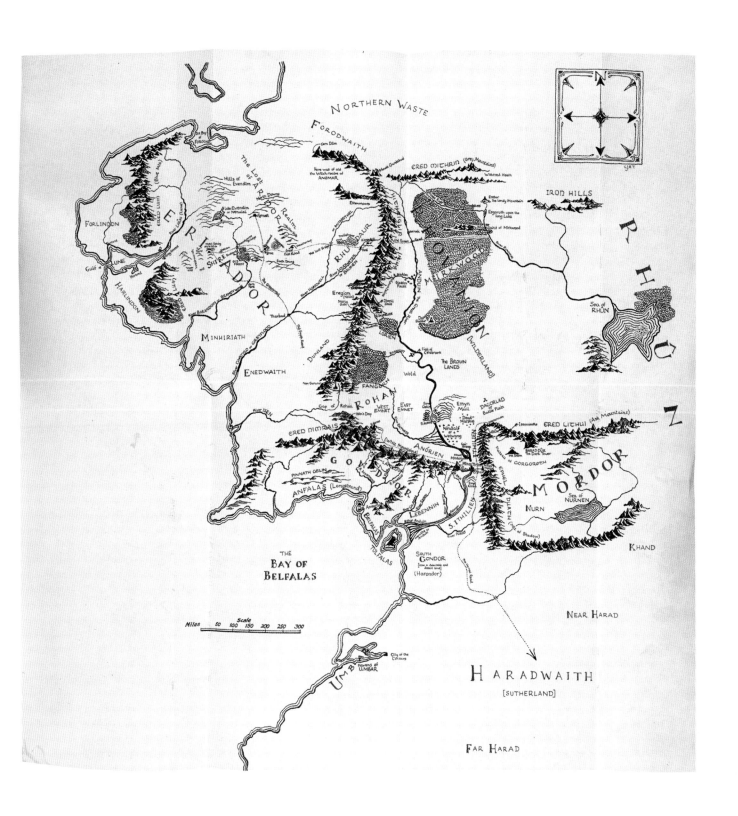

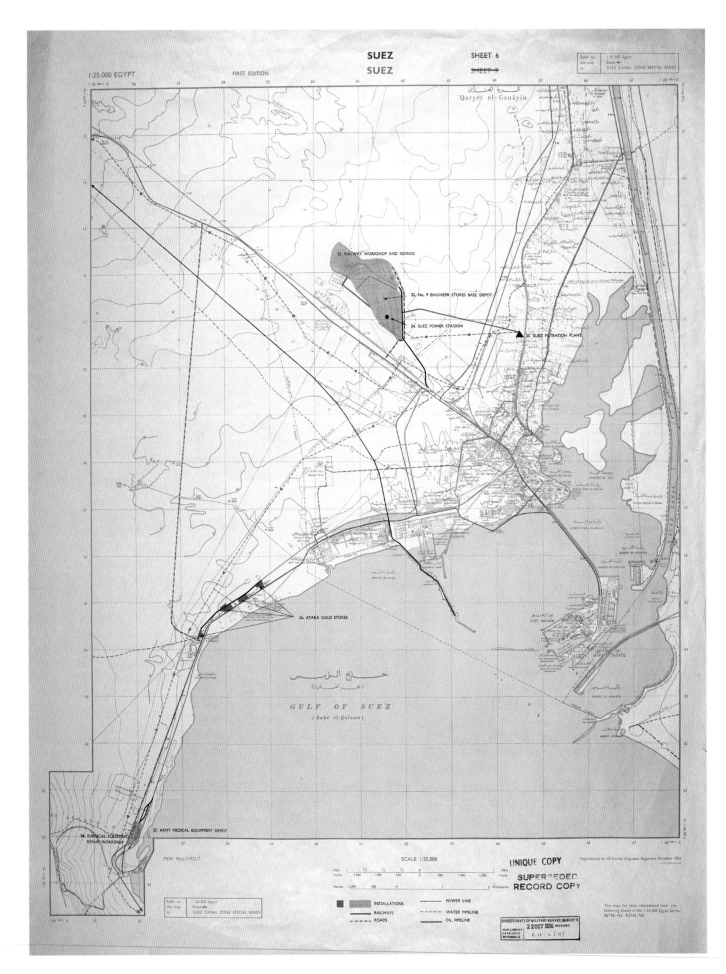

1:25.000 EGYPT

FIRST EDITION

'Qaryet el-Ganâyin

21. RAILWAY WORKSHOP AND SIDINGS

22. No. 9 ENGINEER STORES BASE DEPOT

24. SUEZ POWER STATION

25. SUEZ FILTRATION PLANT

SUEZ

26. ATAKA COLD STORES

PORT IBRAHIM

PORT TAUFIQ

GULF OF SUEZ

(Bahr el-Qulzum)

28. SURGICAL EQUIPMENT REPAIR WORKSHOP

27. ARMY MEDICAL EQUIPMENT DEPOT

MDR Misc(1)1932.7

SCALE 1:25.000

UNIQUE COPY

SUPERSEDED

RECORD COPY

Reproduced by 42 Survey Engineer Regiment October 1954

DIRECTORATE OF MILITARY SURVEY, SURVEY 3
2 2 OCT 1956 RECEIVED
MAP LIBRARY CATALOGUE REFERENCE

INSTALLATIONS   POWER LINE
RAILWAYS   WATER PIPELINE
ROADS   OIL PIPELINE

This map has been reproduced from the following sheets of the 1:25,000 Egypt Series 80(759-765), 81(750-765)

1956: Military mapping during the Suez Crisis

O N 29 JULY 1956 the president of Egypt, Colonel Gamal Abdel Nasser, announced the nationalisation of the Suez Canal. A crucial strategic waterway linking the Mediterranean with South Asia and Australia, the Suez Canal had been designated a neutral zone under British protection in 1888. Nasser's act threatened to further weaken Britain's post-colonial status in the region, as well as placing at risk the transport of oil upon which Britain's 'golden age' was constructed. In the end, the reputational damage to Britain caused by her response would weaken her status still further.

Under the guise of an Israeli invasion of the Egyptian territory of Sinai, Britain and France (who shared a vested interest) proposed to attack Egypt on the basis that they were safeguarding the Canal Zone. But on the discovery of this secret plan, made at Sèvres on 22–4 October, Britain's complicity in the Israeli attack was revealed. Nevertheless, Britain vetoed a United Nations resolution and on 31 October strategic bombing raids commenced, followed on 5 November by the landing of British troops around Port Said at the northern entrance to the canal. Britain was humbled into a ceasefire on 7 November by the United States' threats of economic sanctions.

The tense political situation in Egypt and the Canal Zone during the early 1950s is reflected by the increased production of British military maps. In October 1954 the first edition of the Suez Canal Zone special series mapping was produced by the Middle East Directorate Drawing & Reproduction unit, incorporating data from the copious earlier large-scale maps of the region. These sheets, at a scale of 1:25,000, covered the area on either side of the canal. Printed in brown, with a grid, and place names in both British and Arabic, they provided well-needed coverage for operations and intelligence.

The town of Suez itself, at the southern entrance to the canal, was home to the largest British military base in the entire Middle East. This was, alongside Arab anger at the British-aided foundation of Israel in 1948, a source of antagonism towards the new republican government of Egypt, established in 1952 by military coup. The Queen's coronation tour of 1953 had avoided the area altogether (see p.128).

The Canal Zone maps were quickly brought into action upon Nasser's seizure of the Canal Zone. Sheets that reached the Directorate of Military Survey map library on 3 August, five days later, were overprinted in colour with details of strategic features. The Suez sheet identifies installations such as power plants, railway depots and stores, power lines, water pipes, railways, roads and, probably most importantly, oil pipes belonging to the Anglo-Egyptian oil company (jointly owned by Shell and British Petroleum). These were to be defended at all costs. Further examples of the sheets were accessioned on 22 October, a week before the planned attack.

The existence of the Suez Canal Zone special series reflects British concern for the area: indeed, it would be surprising if such a map had not been produced. Britain's military survey continued to provide unparalleled topographical coverage 'for information', which could be used in readiness if and when the opportunity arose.

1957: Just testing: a nuclear blast over Southampton

I N SEPTEMBER 1957 the Ordnance Survey produced a special sheet of its 1:25,000 series map covering Southampton, overprinted with isodose (radiation) contours emanating from two nuclear explosion points. These explosions were hypothetical, of course, and the map was produced exclusively for civil defence planning and training by the Air Ministry and War Office.[89] Yet as a 'what if' visualisation, it is undoubtedly frightening; it is safe to assume that, had it been publicised, the map would have caused a degree of panic among the civilian population.

The map's existence certainly reveals that a nuclear threat or accident was being taken seriously at official level. The government was dealing simultaneously with the complex issue of nuclear weapons testing. Under pressure from anti-nuclear public opinion, Britain was committed to working towards an international test ban treaty. Yet it was also on the way to becoming a nuclear power, for defence as well as prestige: its testing programme, Operation Grapple, occurred that very same year. Nuclear tests provided the data from which such hypothetical maps could be compiled.

Such was Britain's nuclear dilemma. Operation Grapple, which involved the detonation of thermo-nuclear weapons, took place on Christmas Island in the Pacific Ocean. Despite the limited success of the tests, Britain duly clambered to the 'top table' of undeclared nuclear powers. The year after, the Soviet army conducted the largest ever nuclear explosion in Nova Zembla. A test ban treaty was achieved in 1961, the year before the Cuban missile crisis.

Under a climate of animosity to nuclear power (felt also with regard to Britain's nuclear power station programme, see p. 168), the benefits of nuclear testing were comprehensively outstripped by its potential for destruction. The map shows the radiation effects of two explosions. The first, 'A', shows an explosion in the air. The contours are close together, the effects decreasing fairly rapidly and uniformly, although buildings slow its spread westward. The effects of point 'B's, an underwater blast in Southampton's Eastern Docks, are far more unpredictable and widespread. The radiation is carried northwards along the River Itchen. Buildings and topography create depressions where pockets of radiation linger.

Southampton was one of only a few sheets of this series to be produced for the military, because, as a strategic port, it was one of the areas held to be most at risk. It is perhaps not coincidental that the Ordnance Survey was itself based in the town and, despite its essential operations moving to Chessington in Surrey following enemy bombing in December 1940, still held offices there. In 1955 the government opted to move the Southampton offices to Wellingborough, Northamptonshire, citing reasons of preserving its safety in the event of nuclear attack. Protesting against the move in Parliament, the MP for Southampton, Dr King, dryly refuted the motion, remarking that an 'H-bomb blast in London would blow out windows in Birmingham'.[90]

Whether the map supports or disproves this claim is irrelevant. The move was approved, though the Ordnance Survey in the end stayed in Southampton. Yet Southampton's connection with nuclear fallout remained. The GZ (ground zero) of blast 'B' is, in fact, where nuclear submarines would sometimes berth. Years later, Southampton City Council and the Ministry of Defence issued leaflets to everyone resident within a 2-kilometre radius of precisely this point, advising them what to do in the event of a radiation leak ('Go in, Stay in, Tune in').[91] This map, which was reissued a number of times, brilliantly conveys the conundrum of Britain's post-war nuclear history.

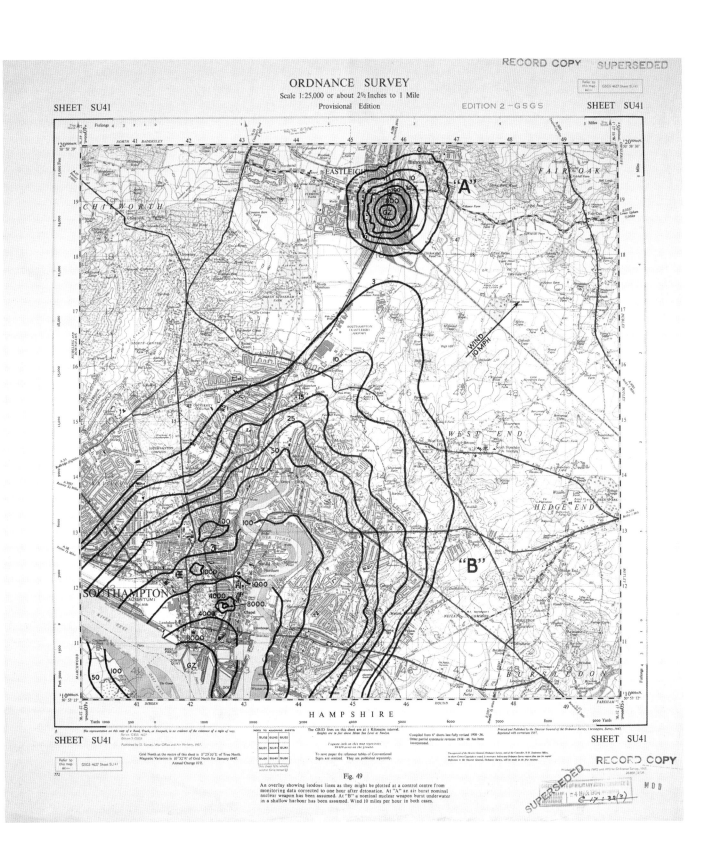

ORDNANCE SURVEY

Scale 1:25,000 or about 2½ Inches to 1 Mile

Provisional Edition

SHEET SU41

EDITION 2 — GSGS

SHEET SU41

Fig. 49

An overlay showing isodose lines as they might be plotted at a control centre from monitoring data corrected to one hour after detonation. At "A" an air burst nominal nuclear weapon has been assumed. At "B" a nominal nuclear weapon burst underwater in a shallow harbour has been assumed. Wind 10 miles per hour in both cases.

1957: A tourist map of Alicante: Spanish holidays and the anticipation of paradise

FROM THE LATE 1950s Spain began to be transformed from a rural, war-torn country into a modern industrialised nation. Its special industry was tourism, and Franco's Nationalist government shaped, controlled and marketed it with sophistication through language, literature and images – including maps.

This folding map of the ancient coastal town of Alicante was produced for British tourists in 1957 by the Junta Provincial del Turismo just when the wider stretch of the south-eastern Spanish coast was on the cusp of being 'discovered' as the Costa Blanca. The first package holidays from Britain began at precisely this time, thanks to cheaper air travel and higher wages. The Costa Blanca, translated literally as the 'White coast', was invented by British European Airways when it launched its route to Valencia in 1957. In 1959 there were 2.8 million visitors (half of them British); this rose to 19 million ten years later.

The British were attracted by the image of an unspoilt (and very cheap) paradise, sunny weather and golden beaches conveyed through state-produced literature and brochures. The Alicante map was one such device, issued free to visitors and those intending to visit, collectable from the offices of English travel agents. Since its aim was to attract visitors, it is drawn in an elegant and – most importantly – 'Spanish' style. It is practical in that it shows the layout of streets and tourist offices, but it is entertaining too, conveying a sense of discovery and adventure, with features such as the steam engine playfully bouncing along the track to Murcia. The street plan is fleshed out with photographs of the beach, castello and harbour, and informative text about Alicante's history, culture and discoverability. This text was printable in German or Swedish.

Of all these elements it is the vast stylised sun that reinforces the principal draw for the British tourist: good weather. The motif, which is still used today by the Spanish tourist board, has become so quintessentially Spanish to prospective holidaymakers that no amount of bad weather experience is likely to dislodge it. The artwork on the cover is by a local artist, Javier Soler, after whom a thoroughfare in the town would be named. The street is not yet on the map, because in 1957 Alicante had not swollen to the extent that street names glorifying famous sons were needed.

The success of Spain transformed the character and make-up of large areas of the Spanish Mediterranean coast. The Costa Blanca was transformed from sleepy coast to bustling resorts of white concrete hotels. Just up the coast from Alicante, Benidorm had been a small fishing town until the mid-1950s, when it began to be transformed into the large holiday town now synonymous with British football shirts, sunburn and alcohol. Unlike Benidorm, Alicante had a distinguished history. It was an important port, which had appeared on sea charts (known as 'portolan charts') since at least the thirteenth century, built on the site of a Roman town, with more genuine sights for the tourist to visit (it had also been the last Republican bastion to fall to Franco in 1939: see p. 90).

Although the town had been in existence for centuries, its 'discovery' by holidaymakers was provoked by the offering up of its unspoilt, authentic nature through maps and images, which continued even after this reality had ceased to exist. Areas like the Costa Blanca adopted the English language, British customs and commodities. The effects were akin to globalisation and colonialism.

1958: The world we live in: overpopulation and uninhabited wastes

HE WORLD WE live in is a world of rapid change, and nobody need be surprised to be told that any map of it is likely to be out-of-date before the ink is dry on the paper.' So wrote the eminent British geographer David Linton in his essay accompanying this dramatic world map, published for the Barclays Group of Banks in 1958. Taken together, the map and accompanying text, tables and images constitute a rallying call for the people of the world to unite, but against what? Against themselves, of course. This map is all about population increase, and the capacity of the world to sustain the escalating rise, which even Linton's statistics (a projected figure of 4 billion people by 1980) turned out to underestimate. In order to find out exactly what this had to do with Barclays Bank, we need to look a little closer.

The map itself, drawn using Bartholomew's 'Atlantis' projection which gives greater prominence to the polar regions, is actually a population density map, the white areas indicating heavily populated areas (an average of 256 people per mile), the darkest areas uninhabited. Yellow dots are cities of more than 1 million people, which are also listed in the map's margins. Linton refers to them as the places

where 'the calamities of flood, storm, epidemic, disease, famine, unemployment and war fall most heavily'. The map's red cover shows multiplying humans emerging from the heat.

In contrast to these urban conglomerations, it is the inhospitable parts of the world that occupy a curious centre stage. The great dry deserts of '15 million almost useless square miles' or cold deserts of 10 million miles, with their attendant racial stereotypes – Antarctica, the Sahara, Siberia – are given huge prominence by how centrally they are positioned in the frame, made to appear accessible, explorable and exploitable. Somewhere in this new proximity seemed to lie at least part of the solution to the conundrum of a multiplying population: science.

It was International Geophysical Year in 1958, and both map and text mark the national research bases on the Antarctic continent, together with the route of Edmund Hillary's successful Commonwealth Trans-Antarctic expedition (see Shackleton's earlier attempt, p. 44). The prominence of Antarctica on the map draws attention to these gestures of international togetherness. Cold War positioning and territorial expansion into the unspoilt continent, as well as space, is scarcely masked. But the 'one world' principle of the map has at least some credibility. From the perspective of Britain and Barclays, which was at the time looking to expand through acquisition into foreign markets, this image can have had only positive effects.

The 1951 Festival of Britain had promoted science and technology as a common ideal and a trigger for change (see p. 124). When in 1958 the Russian Sputnik incinerated on re-entry into Earth's atmosphere it must have seemed that science held the answer to everything (Linton's text includes mention of the first direct flights between London and Moscow). The scientists in Antarctica studied climate change and mineral deposits, which it was hoped would lead to the gathering of raw materials and the development of trade and industry. The course of the latter would be eased by the increasingly small size of the human world thanks to air travel and the then only embryonic offshore manufacturing, trading and banking system (see 'Islands of the Blessed', p. 182). Manufacture and resource could be brought together as never before, saving the human race along the way.

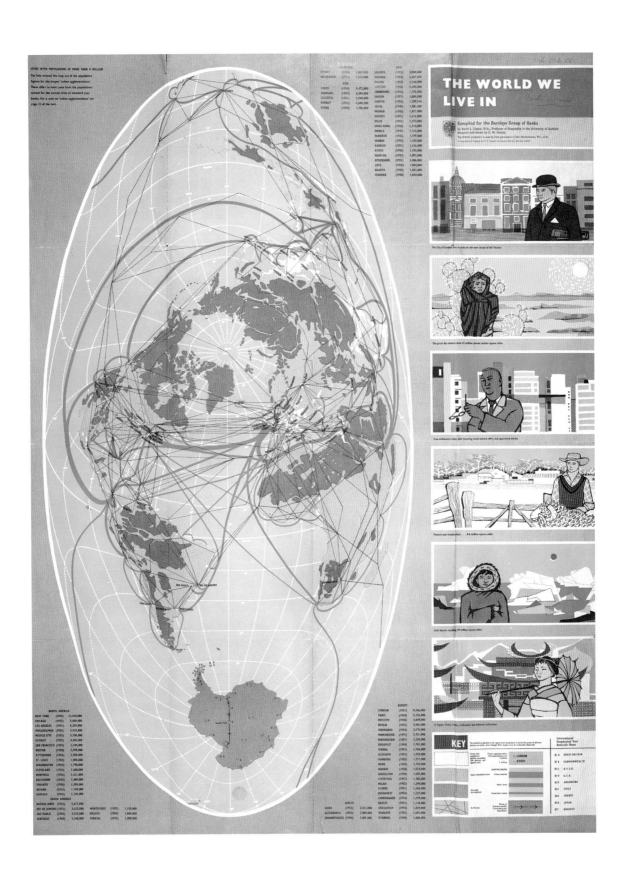

香 港 及 新 界

HONG KONG AND THE NEW TERRITORIES.

Scale 1:80,000

Geographical Section, General Staff, No. 3961.
Published by the War Office, 1936.

Woodland.	Scrub.	Rough Grassland and scrub under 12 inches.	Badlands, heavily eroded.	Arable.	Swamp.	Houses with gardens.	Built... Agric...

THIRD EDITION

Refer to this map as G.S.G.S. 3961
HONG KONG. North Sheet.
Third Edition

NORTH SHEET.

1958: Land use and the creation of modern Hong Kong

HONG KONG WAS ceded to the British in three phases. The island of Hong Kong itself was followed by the Kowloon Peninsular after the First and Second Opium Wars in 1842 and 1860 respectively (see p. 32). Almost half a century later, in 1898, Britain acquired the New Territories on a 99-year lease – a further 365 square miles that gave Hong Kong the water, agricultural land and space to make it viable. It was this lease that shaped both our map and the history of Hong Kong in the twentieth century, predicated on the 1997 handover back to China.

The base map is the standard GSGS (War Office) map of the colony, revised to 1957. It has been overprinted with a land utilisation survey compiled by Thomas Tregear, a lecturer at the University of Hong Kong. It illustrated a detailed report on the subject, edited by British geographer L. Dudley Stamp, which was to be the first monograph of the World Land Use Survey.

The inspiration was Stamp's Land Utilisation Survey of Great Britain, begun in the 1930s, which aimed to provide the most accurate survey of the use made of land since the Domesday Book. Funding was precarious and Stamp enlisted the help of volunteers, including schoolchildren, but the project proved its worth in the Second World War. In 1949 a World Land Use Survey was promoted by the UNESCO-backed International Geographical Union. It seems to have foundered over difficulties in standardising international agricultural land use, but 1950s Hong Kong was undergoing a moment of demographic crisis which threatened its political stability, and successful land utilisation was a key component of the solution which led to the birth of the modern city.

The map shows the built-up areas of Victoria and Kowloon, and the arable areas (blue), mostly in the New Territories. Much of the region is still scrub and rough grassland. The red denotes badlands, which were heavily eroded, and other 'agriculturally unproductive' areas, including those covered by 'camps, cemeteries etc'.

These 'camps' mushroomed as refugees fled the Chinese civil war between the quasi-fascist nationalists and the communists. The immediate impact of this influx of manpower, capital and expertise was overcrowding, water shortages and riots, but the shanty towns swiftly gave way to high-rise blocks. Tregear noted that water was 'so severely rationed as to be turned on for only three hours every alternate day'.[92] However, this combination of factors also facilitated a massive expansion of Hong Kong's industrial base which in turn funded investment in housing, healthcare, infrastructure and education, causing a sharp rise in the standard of living.

Tregear's calls for 'optimum' land use were part of a wider programme of planning undertaken by the British authorities. However, there was no accompanying constitutional reform to edge Hong Kong towards a democratic future: maintenance of the status quo suited all parties in the Cold War era. Fear of offending the Chinese, and the underlying assumption that Hong Kong could never become an independent state, preserved an ossified colonial structure that did nothing to quench the entrepreneurial spirit of the population. Hong Kong today remains a beacon of capitalism as a Special Administrative Region within the People's Republic of China.

1961: Bahía de Cochinos: a quiet little spot for an invasion

THIS MAP OF part of the southern coast of Cuba includes the Bahía de Cochinos (Bay of Pigs), designated as the seaborne landing site of a 1,300-strong militia of Cuban political exiles on 17 April 1961. 'Brigade 2506' had been recruited in Miami, covertly trained in Central America, and armed and equipped for attack by the CIA. The intention of this undercover operation was the overthrow of the republican and anti-US government of Cuba led by Fidel Castro, whose Cuban revolution of 1956–9 had deposed the right-wing presidency of Fulgencio Batista and exiled those now recruited by the CIA.

Cuba's area and waters had been extensively mapped by its Spanish rulers and American neighbours, most recently in the American Geographical Society of New York's map of Hispanic America (1920–45). By the early 1960s, in response to threats from its mighty neighbour and with assistance from its equally powerful ally, the Soviet Union, it was able to map itself. This Instituto Cubano de Geodesia y Cartografia map of the Cuban Republic was produced in 1961, the year of the United States' invasion, at 1:250,000. It was available to a variety of international agencies, including the UK Ministry of Defence, which acquired this set upon its completion in 1965. Britain was just one of many spectators with eyes upon Cuba. How its position in world affairs had changed.

The failure of the invasion is shrouded in controversy. The United States did not want to be viewed by the international community as the aggressor, still less to be associated with such a risky operation, preferring instead to attack hidden behind the guise of counter-revolution. Secrecy would perfectly explain the choice of the Bahía de Cochinos as the suitable landing site: the map shows it to be a quiet area of coastline surrounded by marshland. Importantly, there are airstrips at Playa Girón. Perhaps equally importantly, it was 'hidden' from international eyes to the north and east, and inconspicuous on the map.

The attack came from Guatemala, to the south-west – the opposite direction to the United States. Landing at Playa Girón and Playa Largo at the head of the bay, the attack achieved a beach head as far north as the village of Palpite. But by 19 April, isolated and in the face of armed resistance, the attackers fled, surrendered or were killed. Unfortunately for the effort, Cuban air and naval defences had not been suppressed in the build-up. The element of surprise had not been achieved, with talk of a mooted attack circulating up to two weeks beforehand. No local support was rallied, since in this quiet area there were few people to rally.

When the Cuban army arrived, led by Castro himself, no support for the invaders came from the USS *Essex* that was sitting in the bay. Perhaps support was what the CIA chief John Dulles had hoped for. Later he wrote of an assumption that 'any action required for success would be authorised, rather than permit the enterprise to fail'.[93] President Kennedy's decision not to employ these troops may again have been based on the United States' wish to retain 'plausible deniability' of the operation, but in the end it contributed to comprehensive failure.

Castro's Cuba remained a thorn in the United States' side, 90 miles off its own coast. After the revolution Castro had nationalised US assets, including oil refineries, the printing house of the *Reader's Digest* and a toothpaste factory. His political stance – and his links with the Soviet Union – stood in direct contravention of the 1823 Monroe Doctrine, which had claimed the entirety of America as beyond Europe's sphere of influence. The botched invasion not only popularised Latin American opposition to the United States, but pushed a wary Cuba closer to the Soviet Union.

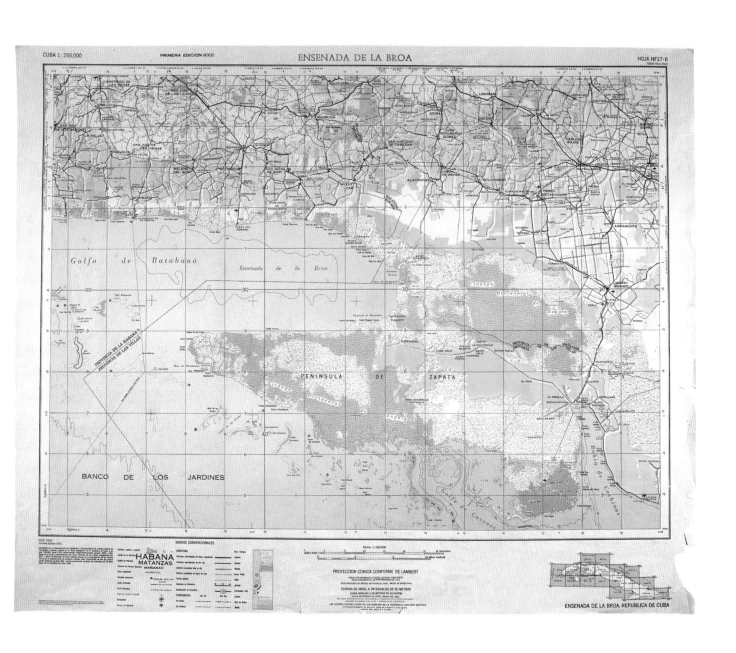

The European Community

INDUSTRY I : Selected industries

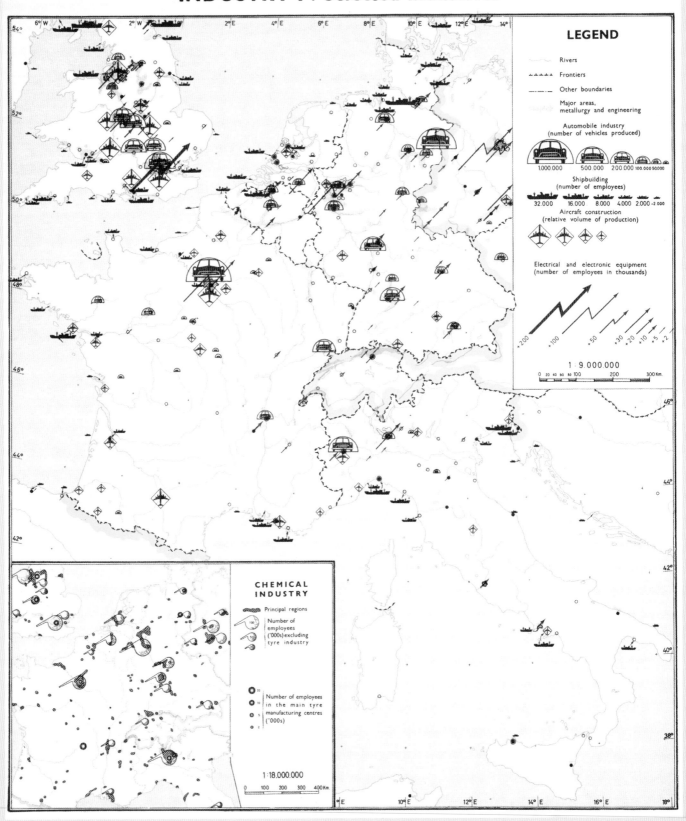

LEGEND

Rivers

Frontiers

Other boundaries

Major areas, metallurgy and engineering

Automobile industry
(number of vehicles produced)

1.000.000 500.000 200.000 100.000 50.000

Shipbuilding
(number of employees)

32.000 16.000 8.000 4.000 2.000 -2.000

Aircraft construction
(relative volume of production)

Electrical and electronic equipment
(number of employees in thousands)

+200 +100 +50 +30 +20 +10 +5 +2

1 : 9.000.000

0 20 40 60 80 100   200   300 Km.

CHEMICAL INDUSTRY

Principal regions

Number of employees ('000s) excluding tyre industry

Number of employees in the main tyre manufacturing centres ('000s)

1 : 18.000.000

0 100 200 300 400 Km.

1962: 'The European Community in Maps': preparing the way for the UK

THE EUROPEAN ECONOMIC Community (the origin of today's European Union) was based on a set of economic and legal agreements, ratified by the Treaty of Rome in 1957; the signatories were six European countries, including France and West Germany, but not the United Kingdom. It created conditions amenable to free trade for those nations wedged in between the communist-run countries to the East and the United States. Between 1961 and 1963 a new round of membership applications was solicited, but in 1963 Britain's application to join was turned down.

'The European Community in Maps' was produced in 1962 for the purposes of education while Britain's application was in process. The eleven maps showed population, industry, transport networks, nuclear power and industries. Despite not being a member, Britain (though just England and Wales are shown) is included in the maps and accompanying tables. These show it to be the strongest, most populous and industrious part of Europe, equal with the Ruhr region in Germany, London vying with Paris. Written in English, there is no doubt as to whom the publication was intended to persuade of the value of Britain's joining the EEC. Economically, it may have been an attractive option to many in Parliament, but Britain itself was divided and more than a little 'Eurosceptic'.

The maps may also have been intended to show Britain as a 'natural' EEC member for Anglophobe Europeans. One of these was the French president Charles de Gaulle, who in his veto speech in 1963 pointed out Britain's insular nature and the fact that, in the Commonwealth, it did actually already have its own trade organisation. In addition, Britain was close – too close in de Gaulle's eyes – to America. Britain's ploy to offer France a connection with US expertise over developing nuclear power was unsuccessful.

Despite the bonds between France and Britain forged in two world conflicts, and to a lesser degree their partnership in the disastrous attempted seizure of the Suez Canal Zone in 1956 (see p. 134), there was no room for both countries around the table in 1961. 'De Gaulle goes back to his distrust and dislike like a dog to his vomit', British prime minister Harold Macmillan is supposed to have uttered upon hearing of the veto.[94]

So well did the 'European Community in Maps' illustrate Britain's strong economic performance against those of existing member states that the average Brit might have been forgiven for wondering what benefits EEC membership and its common agricultural policy could possibly have brought them. When Britain finally gained entry to Europe, along with Denmark, in 1973, the gross national product of Portugal was a mere fraction of Britain's. The thought of subsidising poorer EEC members may well have turned public opinion against European membership, as indeed it has continued to do to this day.

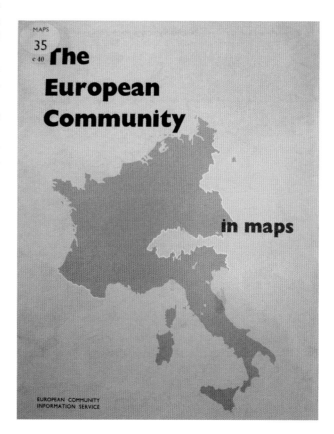

THE TIMES MAP FOR THE GENERAL ELECTION

1964

SCOTLAND & NORTHERN IRELAND
FOR AREA 1b SEE SCOTTISH LOWLANDS INSET

1a
SOUTH LANCASHIRE & S.W. YORKSHIRE

2a
LONDON & SUBURBS

3a
NORTH-EAST INDUSTRIAL

4a MIDLANDS (CENTRAL)

1964: A do-it-yourself general election map

OR EVERY ONE of the UK's parliamentary, local and European elections held during the course of the twentieth century, maps have been made available to show the geographical pattern of results. This map was published before the 1964 general election, uncoloured so that the owner could add in the results themselves, enabling them to feel a part of this important constitutional event. Although the first televised election was in 1959, televised coverage really came into its own in 1964, by which time far more people had television sets. As a result, the 1964 election was the first really populist election.[95]

The electoral map of the UK had changed dramatically over the nineteenth century as boundaries were altered and constituencies created to better represent a population that had both increased and concentrated in large urban, industrial, working-class areas. That a tiny segment of Glasgow had the same parliamentary value as the vast Scottish Highlands area of Ross and Cromarty illustrates the problem. It wasn't until the 1990s that advances in technology would combine with increased boldness to more properly represent these complexities (see the *New Social Atlas of Britain*, p. 212). Inset enlargements sufficed here for the more densely represented places.

The general election of 15 October 1964 was also one of the closest fought in British history. Harold Wilson's opposition Labour party defeated the Conservative government under Alec Douglas-Home, taking just a five-seat majority to the House of Commons. The third party, the Liberals, were returned in nine seats. The victory put Labour, the party striving to become associated with modernism and change, into power for the first time since 1950. Wilson's 'white heat of revolution' speech in 1964 emphasised the technological, industrial drive which found cartographic expression in the *Atlas of Britain* the previous year. Two years later another election saw Labour win an increased majority.

This map was produced by *The Times* newspaper, but apart from the modern bright orange background (polychrome printing was still uncommon for inserts – see the 1971 football map, p. 156), it is left for the reader to supply the colouring. Looking closely, the owner of this example has not adhered to the boundaries, colouring large swathes – suggesting that they did so at the end of the night when all the results were to hand. The pattern of the electorate may also have helped, with uninterrupted swathes of Conservative heartlands in the south and borders, separated from the overwhelming red of the inner cities, Labour's traditional working-class strongholds.

Souvenir participatory maps to cover all public occasions had become widespread. In 1966 numerous maps and charts were produced to accompany the World Cup, held in England that year. This map denotes a populist, youth-oriented edge creeping in even to politics.

1965: Glasgow residential land use: housing solutions for 'Empire's second city'

SOLUTIONS TO GLASGOW'S post-war housing problems met various fates. The relocation new towns of East Kilbride and Cumbernauld flourished, while residential tower blocks such as Red Road Flats, at one time the tallest in Europe, are demolished or condemned. In the 1960s the inadequate, insufficient and bomb-damaged inner area of 'Empire's second city' had been flagged up not only by local and national governments, but also university geography departments.

The context of this map is not a governmental report or council proposal, but an article in the *Geographical Journal* of 1968 entitled 'Topographic science', which described the types of geographical study carried out in universities. Produced by Michael Wood, a geography student at Glasgow University, it is not so much a land use map as a map of housing types. The residential areas of Glasgow city are coloured to distinguish cottages, tenement housing, terraces and flats, the darkness of tone indicating the age of each area.

Drawn from survey data, the map provides the history of Glasgow's housing at a single glance. The most notorious, the tenement blocks, were inner-city solutions to the large influx of rural workers from the 1880s; by 1945 they were dilapidated and overcrowded.[96] Terraced housing, so prevalent in other British cities, is confined mainly to the north-east of Glasgow. Much of it was damaged by German bombing and replaced by cheaper tenement housing. Cottages, the expensive interwar housing solution, correspond with the areas to the north-west and south, which came under the city's jurisdiction in 1925 and 1938. By the mid-1960s, however, the solution to overcrowding and its attendant social ills was the tower block. These are indicated by small grey areas bordered in black. The large complex of ten-storey flats at Moss Heights is shown, while in 1965 work had just begun on the eight Red Road Flats in the north-east of the city.

Like much of the 1950s tenement housing, these were located not in the city centre but the surrounding green-belt area – yet crucially they were still within the city's jurisdiction.[97] The government's Clyde Valley report of 1946 had proposed outsourcing population and industry to the new towns. Yet the city council wished to retain business and taxable residents, not to mention any regeneration funding, inside its boundary. A particular civic pride in addressing its own problems is also detectable: along with poor inner-city social conditions, the decline of the core industries of great Glasgow had brought with it strong working-class civic politics.

This was precisely the sort of socio-scientific subject that geographers explored through statistics and skilfully interpreted through the visual language of maps. Presented in such a way, the basis for and solutions to Glasgow's problems emerged. Scientific confidence through the study of geography was one aspect of modern Britain. Another, of course, the tower block, was considered a great civic achievement,[98] the demise of which signalled the end of the modern age. The great modernist social housing project Pruitt Igoe, built in St Louis, Missouri in 1951, was demolished in 1972. Glasgow's Red Roads housed, among other minorities, Kosovo refugees during the 1990s, before their demolition began in 2012.

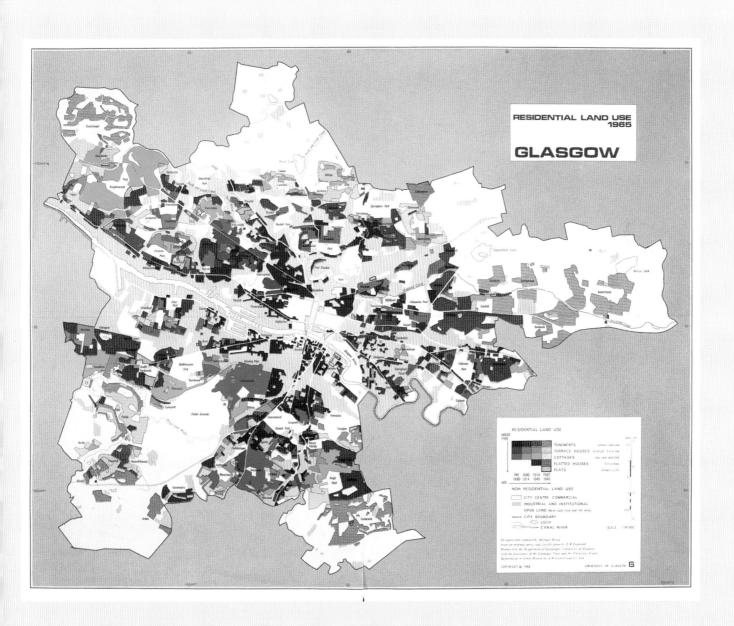

RESIDENTIAL LAND USE
1965

GLASGOW

RESIDENTIAL LAND USE

HOUSE
TYPE

TENEMENTS
TERRACE HOUSES
COTTAGES
FLATTED HOUSES
FLATS

PRE 1880 1914 POST
1880 1914 1945 1945
AGE

NON RESIDENTIAL LAND USE

CITY CENTRE COMMERCIAL
INDUSTRIAL AND INSTITUTIONAL
OPEN LAND
CITY BOUNDARY
LOCH
CANAL RIVER   SCALE 1:50,000

COPYRIGHT © 1968   UNIVERSITY OF GLASGOW G

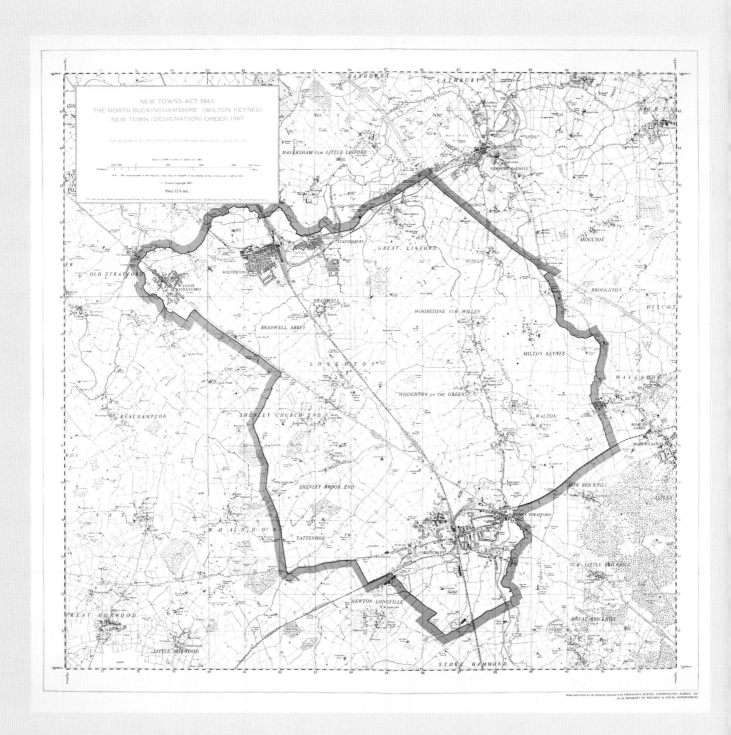

NEW TOWNS ACT 1965
THE NORTH BUCKINGHAMSHIRE (MILTON KEYNES)
NEW TOWN (DESIGNATION) ORDER 1967

C Crown Copyright 1967

Price 12·6 net.

1967: A map of the proposed new town of Milton Keynes: the spirit of the age

THE THICK BLUE border lithographed on to this Ordnance Survey 1:25,000 sheet outlines the 34 square miles of north Buckinghamshire countryside designated for one of Britain's new towns. Milton Keynes – its name taken from one of the many villages subsequently swallowed up by it – was an urban development resulting from the New Towns Acts of 1946 and 1965, borne of a need to provide overspill housing primarily for overcrowded London. Its design was strongly influenced by Ebenezer Howard's revolutionary garden city movement of the late nineteenth century. This preached a greater synergy of urban and rural elements in town planning, motivated by social, moral and public health dimensions – perfectly in tune with late 1960s social ideology.

The map was produced for the Department of Housing and Local Government to accompany the Milton Keynes Designation Order. This piece of legislation removed the area from local government control and placed it under the jurisdiction of the newly appointed Milton Keynes Development Corporation. This map presents the initial legal step from which compulsory land purchase orders would follow, making an unequivocal statement of responsibility. The definition of the area using blue, a purposefully passive, neutral colour, for a boundary which nevertheless connotes inclusion and exclusion, loss (and gain) of authority, provides an interesting perspective upon the ethos behind the town's planning and what the town has become.

Most British towns developed organically over hundreds of years, spreading along major arteries as strip development and reaching beyond borders as a result of natural forces. Marking out a town's limits prior to its development is therefore something rather extraordinary to Britain, if not remotely unusual in the case of the generally younger towns of North America. However, drawing a line around some land is certainly nothing new; the British countryside had itself long been divided up and apportioned into fenced and hedged fields with angular copses. Not only was Milton Keynes no different from the historically man-managed British countryside, its design yielded to it too.

The border suggests a division between the urban and the rural, yet the town was designed with 20 per cent parkland and millions of trees, which would grow to obscure the major roads and low-height, low-density housing. The grid design, along North American urban lines, was in fact more of a mesh through which would grow what was below. Designed as a 'lazy' grid design which 'responds to topography and the accidents of geography',[99] it followed the natural slope of the land (as opposed to Medicine Hat, see p. 38). Far from being accidental, the shape corresponded with the terrain of the land which extended beyond the blue border in all directions.

Although Milton Keynes was a product of the 1960s spirit of modernity, it would subsequently suffer derision as a city devoid of

any 'history'. The confident creativity of the architects and planners of the Milton Keynes Development Corporation proceeded to look out of place during the depressed periods of the subsequent decades, and the entire project was susceptible to periods of economic downturn. The corporation was terminated in 1992, and the blueprint of the 1960s would have to withstand tinkering due to altered planning laws rushed in to meet further housing crises of the sort that had originally made Milton Keynes possible. It would have to deal with unanticipated issues such as antagonism between the resident population and traveller communities, attracted to the city by the plentiful open green spaces.

Such was the gap between the plan and the reality. Yet the confident authenticity of Milton Keynes may be put into perspective by the example of another new town, that of Poundbury in Dorset, which was constructed on the Duchy of Cornwall estate in the 1990s. Poundbury was built as a replica of the British country town, with high-density housing that was both architecturally homogeneous and faithful in details such as the bricked-up windows, aping window-tax-avoidance measures from the late seventeenth century onwards. As a brave statement of the confident, forward thinking achievable in Britain in the 1960s, one need look no further than Milton Keynes.

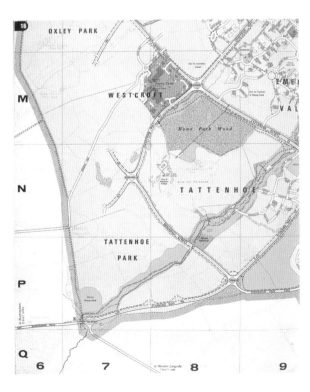

1968: Earthly concerns:
a photograph of the Earth and a medal of the moon

TWO RELATED MAP images reflect the human concerns embedded in the new and profound experience of space exploration. The first is an important photograph showing the Earth from space, taken during the United States' Apollo space programme (1961–72). Initiated by John F. Kennedy in 1961, the programme's aim was to put a man on the moon by the end of the 1960s.

Although the established goal may have been the moon, one particularly interesting and unanticipated by-product was the over-the-shoulder glimpse back at Earth. This viewpoint, which Denis Cosgrove called 'Apollonian perspective, so long anticipated in imagination',[100] was captured in a number of photographs taken on board Apollo 8 as it orbited the moon on 24 December 1968. This one, 'Earthrise', was taken by the astronaut Bill Anders and shows the Earth rising up from behind the grey surface of the moon in front of an impenetrable black background.

The photograph appeared in newspapers and magazines across the world, acquiring a number of meanings and coming to symbolise 'earthly' issues, some quite removed from its original one. Because the Earth looked small, alone and vulnerable in a void, uniting and optimistic messages were evoked, defined as 'whole world' and 'one earth' discourses.[101] These saw 'Earthrise' and its accompanying photographs incorporated into logos, and appropriated for causes such as environmentalism and nuclear disarmament (the more famous 'Blue Marble' photograph together with its ecological connotations would later be appropriated by Google).

Despite these implied meanings, the context of the space programme remained that of international Cold War politics. Space, like Antarctica (see p. 140), offered a place into which humans might expand without causing friction with neighbours.[102] The photograph may have come to signify world togetherness, but it also represented an American triumph in the ongoing ideological war against the Soviet Union.

NASA landed a man on the moon on 20 July 1969, and the second map illustrated here was produced in order to commemorate the event. It is a silver medal, one face bearing the inscription 'Presented by the Royal Geographical Society, First landing on the moon, 20th July 1969 – to Neil Armstrong, Edwin Aldrin, Michael Collins'. On the other face is a relief map of the moon, based on a drawing by the RGS Chief Draftsman G. S. Holland. Extending to the rim, the detailed topographical description of the moon's surface includes the name of the Sea of Tranquility and marks the landing spot with a cross.

As an organisation established to encourage and support exploration, the RGS quite rightly wished to commemorate the feat. But this medal doubtless also represented a desire to be associated with this particularly *Western* achievement through shared scientific and ideological values. However, in contrast to the bewildering financial cost of American and Russian space programmes, the small-scale geopolitical positioning of the RGS remained subject to economic limits. A quotation for producing the medal by the Royal Mint (£169 for 22-carat gold, £20 10 shillings for silver) was rejected by the RGS as too expensive.[103]

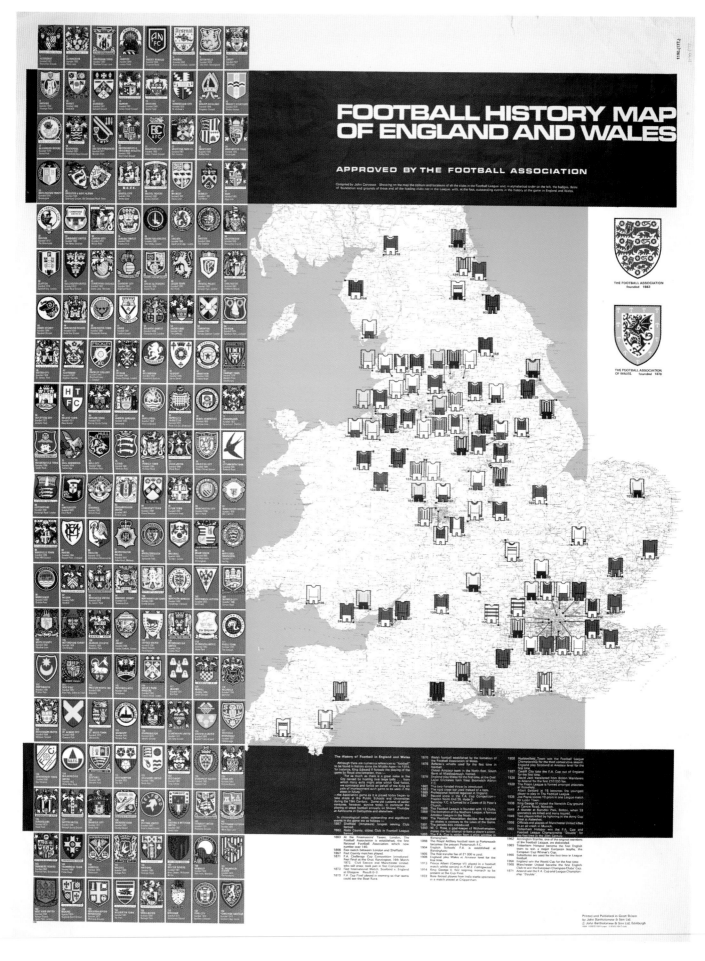

1971: Heritage in glorious colour: a football history map of England and Wales

THE WEEKLY FOOTBALL magazine *Shoot!* was launched in 1969, three years after England had won the World Cup. Containing fast-paced stories, player profiles, glossy colour photos and action shots, it provided a counterpoint to more serious football coverage on newspaper back pages. This youth-oriented magazine tapped into the enormous popularity of association football among the adolescent (mostly male) population of Britain, many of whom played football in the street or in parks and went with their dads to watch professional games. Enthusiasm for football transcended both age and, with the World Cup triumph of 1966 against West Germany, the game's traditionally working-class audience.

Action photographs of 1966 heroes Bobby Moore and Alan Ball adorned the cover of the *Shoot! Annual* of 1971. In the same year, the Edinburgh map publisher John Bartholomew and Son published the *Football History Map of England and Wales*. Costing 50 p, this large 'illustrated map in full colour' included the shirt colours (positioned over the map), crests and histories of each of the 159 league clubs, as well as a selective timeline of the English game's notable events. Apart from the cover illustration, the map's vision of the football heritage of England and Wales appears desperately reserved compared to *Shoot!*'s 'action-packed pages all in colour'. Yet memories of the map are vivid and enthusiastic, the blogger Peter Miles writing how 'the folded paper map would reveal a hitherto unparalleled cornucopia of information, veritably awash with colour'.[104]

In 1971, colour television was still new and exciting, emerging as it was from a world of black and white. The first football match to be televised in colour was broadcast on 15 November 1969, and widespread printed colour was still a novelty too. While comparatively muted, the colour of Bartholomew's map still offered a more vivid appearance than the newspapers and that statistical staple, the *Rothmans Football Yearbook*. It presented football's heritage in an exciting, accessible format, a trainspotter's guide to the beautiful game.

'I had the exact same map on my bedroom wall in Frobisher Green during the 70s', writes another former owner online, and 'it taught me a lot about the geography of England';[105] while Miles adds 'it was surely the catalyst for discovery and spent many years on my childhood bedroom wall, apologies Aldershot for the pin hole'.[106]

The map, as with the game of football, contained a number of different cultural, social and historical features. 'History' is not confined to the timeline. The club crests comprise a mixture of heraldry, civic and industrial symbols, with a few modernisations. The oldest clubs were then moving close to their centenaries. The Football Association (FA) itself, which had prominently endorsed the map (and may well have had a say in its content), had formed as early as 1868. The image of a shared football heritage and collective identity presented on the map fitted conveniently with the FA pursuit of a level playing field.[107] Close and balanced competition offered maximum excitement for viewers, thus increasing English football's commercial attractiveness and marketability (lucrative television coverage deals would not transform the game until the 1990s). The concept of 'shared identity' certainly seems at odds with the aim of putting as many goals past one's rivals as possible, as well as the natural affection of rival supporters of clubs such as Cardiff and Swansea for each other.

The concentration of football clubs largely in working-class urban areas would remain a visual marker of the sport's roots, an indicator through the map of the population density of England and Wales (see the *New Social Atlas of Britain*, p. 212). These were the areas hit hardest by the industrial decline and depression of the mid-1970s onwards. The game and its culture had already reached a golden age, but it was a game on the edge of innocence. The map showed football in colour, but a darker era, which today appears to have been punctuated by hooliganism and racism on the terraces, lay just ahead. But regardless, for those with football posters, scarves and *Shoot!*, football remained the main reason for living.

1971: At face value: the map in conceptual art

N ORDNANCE SURVEY tourist map of Dartmoor forms part of this conceptual work of art by the Bristol-born artist Richard Long entitled *A Hundred Mile Walk*. The map shows the area of land through which Long walked 100 miles in 1971–2. Long has drawn the course of his walk on to the map, included a photograph of part of the landscape, and produced a short sentence for each of the seven days he spent walking. Maps had existed in public art galleries for much of the twentieth century (and far earlier in private galleries of monarchs) as 'ready-made' objects, collage material, and symbolic reference points in drawings and paintings. Quite unusually, Long is here using the map at face value for its representation of a place.

This is termed 'Land art' – varied artwork created outside of the restrictive (even corrupting) confines of the gallery space. Land art developed in the United States at the end of the 1960s as part of a response to what was seen as over-theorised Minimalist art, alongside the more material-based Arte Povera movement. It was back-to-basic art, part of the cultural blossoming or mellowing after the frenetic 1960s. One of the main characteristics of Richard Long's work was the interaction between artist and landscape. In an earlier famous piece, *A Line Made by Walking* (1967), he had photographed a patch of grass on which he had worn a straight line by repeatedly walking over. As with *A Hundred Mile Walk*, the artwork is the walk enacted outside the gallery, not the representational assemblage including the map exhibited in it.

Walking in national parks with the aid of maps is a popular pastime. We've seen how the Ordnance Survey began to capitalise on the market for leisure maps from the 1920s; at the beginning of the 1970s

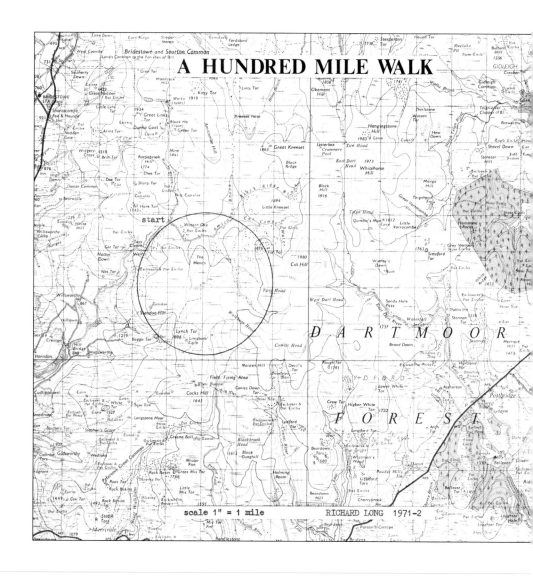

the popularity of outdoor pursuits was increasing alongside a wider concern for nature and the environment. Although in this context *A Hundred Mile Walk* is perfectly understandable, is it appropriate to 'elevate' such conventionality by placing it in an art gallery? This is surely Long's point. The gallery does not elevate, it limits. It is the everyday practice and the experience that is special.

Looking at the map, we see that Long's drawn route is a perfect circle with a circumference of 100 miles. It is a remarkably difficult course to plot and plan, but also to walk, navigating in a giant circle across irregular Dartmoor terrain. Drawn upon the map, it reminds us of the numerous non-natural symbols such as circles, triangles and squares explained in the legends of maps. Similar discourses on the representational limitations of the map were beginning to preoccupy map historians at this same time.

The map is not the artwork, nor the photograph; they merely represent the landscape in which the artwork took place (in other works Long has turned to the land itself, mud and stones, for his material). Together with the lines of simple text, which range from the descriptive ('Winter skyline, a north wind') and the poetic ('In and out the sound of rivers over familiar stepping stones') to the profound ('As though I had never been born'), these are conventional mediums which invoke the geographical imagination of the viewer, though in themselves they are limited surrogates of reality. Long's conventional yet extraordinary response to the landscape reconnects with Romantic art, the landscapes of Turner and the work of the Lake Poets.

Day 1 Winter skyline, a north wind

Day 2 The Earth turns effortlessly under my feet

Day 3 Suck icicles from the grass stems

Day 4 As though I had never been born

Day 5 In and out the sound of rivers over familiar stepping stones

Day 6 Corrina, Corrina

Day 7 Flop down on my back with tiredness
       Stare up at the sky and watch it recede

1972: Four decades on: an attempt at the truth behind Bloody Sunday

O
N 30 JANUARY 1972 thirteen men were shot dead by soldiers of the British Army Parachute Regiment during a civil rights protest in Derry/Londonderry, Northern Ireland. An inquiry into the events of Bloody Sunday one month after the incident exonerated the army. But in 2010, thirty-eight years after the events, another report found the shootings to have been unlawful. Both reports were hampered by their political contexts. The Widgery Report was concluded before the dust had settled, whereas the later Saville Report had to peer back through the mists of time.

After a decade-long investigation, the Saville Report was published on 15 June 2010. On the same day an interactive map appeared on the online version of the *Guardian* newspaper. It consisted of seven animated screens illustrating the course of events of two hours during the afternoon of Bloody Sunday, with extracts from the report and contemporary photographs. In frame two we observe the route of the Loyalist protest march moving through the Derry streets. A further movement is then diverted by army barricades into the exclusively Catholic Bogside area, which morphs into a bird's-eye view. Blue arrows of advancing armoured army vehicles cause the march to scatter in a number of directions. Next we see the 'mopping up' movements of soldiers, and in the final screen the thirteen red dots marking the places where protesters were killed, with photographs of their faces.

The accompanying quote from the report reads thus: 'None of the casualties shot by soldiers of Support Company was armed with a firearm or (with the probable exception of Gerald Donaghey) a bomb of any description. None was posing any threat of causing death or serious injury. In no case was any warning given before soldiers opened fire.'[108] The findings were by no means universally accepted. Much remained contentious, including who had fired the first shot. Yet the neat graphical simplicity of the *Guardian*'s multi-media experience suggests otherwise. Straightforward and clear, it suggests a clarity of understanding of the event despite the intervening time, the consequent gaps in knowledge and the continuing biases.

The *Guardian*'s infographics editor Paddy Allen and graphic artist Paul Scruton produced a number of other virtual reconstructions (including one of the 2011 murders in Norway by Anders Breivik) using a similarly cold and measured presentation. But far from being incidental, emotion, fear and anger were some of the most important features of Bloody Sunday. The movement and noise of the march is not captured, nor the menace of the armoured cars, the violence, or the fear. The simply rendered buildings in the map do not represent their actual dilapidated state and with it the poverty, unemployment and prejudice that had contributed to the Troubles. The wider context of the Catholic minority in Northern Ireland, subsequent incidents of internment and torture by the army, and reprisal murders are not alluded to.

This is because no map can truly show everything (though many others have tried: see, for example, the cartographic chronicle of the fall of Yugoslavia, p. 220). The problem was that the long and divisive legacy of Bloody Sunday had become as much a part of the event as the actions that took place on 30 January. The *Guardian* newspaper had actually sympathised with the army immediately after the event. Without the complete context, a full understanding is impossible.

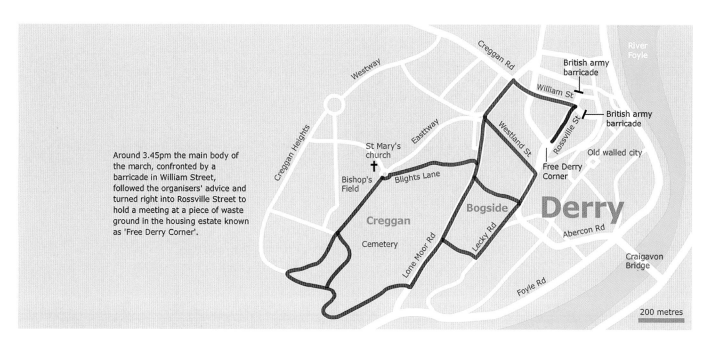

Around 3.45pm the main body of the march, confronted by a barricade in William Street, followed the organisers' advice and turned right into Rossville Street to hold a meeting at a piece of waste ground in the housing estate known as 'Free Derry Corner'.

200 metres

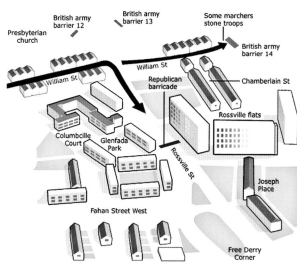

Out of the Red and Into the Blue
1973–1999

The oil crisis of 1973–4 stemmed from the embargo imposed by Middle Eastern oil-producing countries, angry at the United States' support for Israel during the Yom Kippur War of October 1973. With the world economic downturn of the early 1970s, the post-war boom was over, and for the remainder of the decade governments struggled with the resultant social unrest. For Britain the 1970s are characterised as a period of malaise with unease over Europe (Britain was admitted to the EU in 1973), industrial strikes, football hooliganism and a bombing campaign by the Irish Republican Army (IRA). The three-day week and the 1979 'Winter of Discontent' led to a change in the style of government and the end of consensus politics.

The policies of the Conservative Party, led by Margaret Thatcher, improved Britain's economic outlook by introducing free-market economics and privatising those industries that had been nationalised by the post-war Labour government (a policy continued by Tony Blair's Labour government from 1997). Heavy industries long in decline were dismantled completely, resulting in devastating unemployment. The power of trade unions was diminished. High and unpopular taxation policies led to riots. The 1982 Falklands War between Britain and Argentina demonstrated a hard-nosed British nationalism (see p. 184). The United States similarly protected its perceived sphere of influence with military action in Grenada (1983), Nicaragua (1986) and Panama (1989). Dictators such as Chilean Augusto Pinochet found favour and hospitality with Thatcher and Ronald Reagan.

Deterioration in the relations between East and West was underlined by America's involvements in South America, but also by the Soviet invasion of Afghanistan in 1979. The Soviet leader Mikhail Gorbachev attempted reforms of the Soviet political system from 1985, but communism was crumbling. In 1989 the Berlin Wall came down, followed by revolutions in Romania, Czechoslovakia and Poland. In 1991 the Soviet Union itself was broken up. The vacuum this left caused deep instability, particularly in Chechnya and the Balkan region, where the break-up of Yugoslavia led to the Balkan War waged by Serbia upon its neighbouring states and various ethnic groups (see p. 220). Serbia was bombed into submission by United Nations forces in 1999.

In Africa the post-colonial fallout intensified during the 1970s. Military-governed African nations such as Sudan and Nigeria experienced civil war and coups, supplied with arms by both the West and East. Bloodshed followed Zimbabwean independence in 1980 (see p. 176). The Sudanese and Ethiopians were among those worst hit by famine during the early 1980s. The Live Aid rock concert of 1985 raised money for famine victims, prominently showcasing a Western conscience (see p. 192). It also served to demonstrate the technological capability of live global video link-ups, as well as the power of television in the home.

From the 1970s, at the same time as the photocopier entered offices, the seeds of digital revolution were sown. Digital printing gradually replaced photographic reproduction. Digitisation of the Ordnance Survey's archive was complete by 1990, and digital mapping enabled all manner of statistics to be visualised geographically. Information could be beautiful. Arcade games and personal computers became more commonplace, and the public release of the Internet in the early 1990s inaugurated an 'age of information' (see p. 208). Media and rolling twenty-four-hour news was exemplified by the Gulf War of 1990–1, the first 'televised war' (see p. 202). By the time of the terrorist attack upon the World Trade Center in New York in 2001 (p. 228), the media had become a weapon in itself, an important tool of propaganda and publicity.

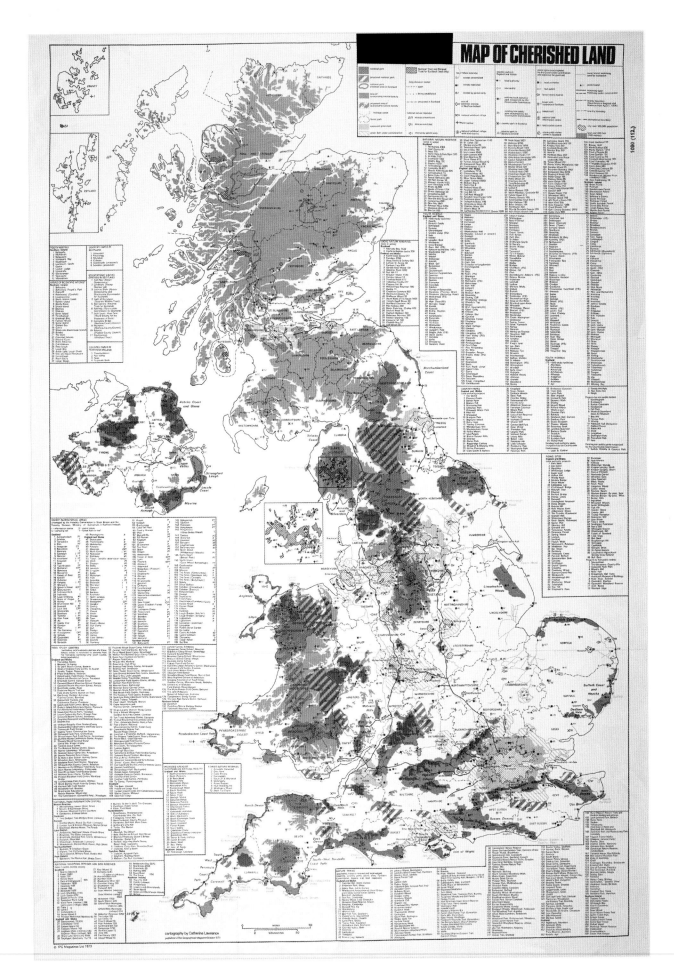

MAP OF CHERISHED LAND

1973: Maps of 'Cherished land' and Industrial expansion

THE *GEOGRAPHICAL MAGAZINE'S* 'Map of Cherished Land' is both a tourist map and a protest map. The two are not mutually exclusive. Published in October 1973, alongside an article entitled 'The cherished land which must be preserved NOW', the map indicates parts of the British Isles that are designated national parks and areas of outstanding natural beauty (AONB), how to reach them, and how to enjoy their picnic sites and nature trails. It also implies that they are endangered not only, perversely, by excessive tourism,[109] but by changes to planning and development laws, and the implications of joining the European Union. Looked at in a wider context, the map is another embodiment of widespread identification with environmental issues beyond those of traditional countryside lovers, tapping into the almost mystical affinity of the British with their 'green and pleasant land'.

By the 1970s, major national environmental groups, such as the National Trust (formed 1895), the Commons Preservation Society (1865) and the Ramblers Association (1935), were mature societies with large, varied and popular membership bases. Their long-standing interests included defence of the 'right to roam', greater prominence of rural affairs in government policy, and preservation of the countryside against the spectres of urbanisation and industry. Legal protection for land existed in the designated status as national park, AONB or green belt. The first AONB, the Gower Peninsula in south Wales, had been created in 1956, and the Countryside Commission (formerly the National Parks Commission) came into being in 1968.

This 1973 map shows Britain as a complicated patchwork of beauty spots, many of which – the Pennines, the Lake District, Snowdonia – have more than one safeguard status. The Cambrian Hills in Wales had, according to the article, just lately been declined the status of national park. Of particular interest, however, are the number of proposed AONBs shown in dashed diagonal lines, thirty-three of them compared with the existing forty-one. Why were so many under proposal in 1973?

The answer lies in the fact that although the environment was a strong theme of the 1970s, so was economic depression, against which the preservation of the environment is always difficult to maintain. Urban development, enabled by the New Towns Acts of 1946 and 1965, eased overcrowding in cities and created phenomenal numbers of jobs. In 1967, the large town of Milton Keynes was projected upon 34 square miles of rural north Buckinghamshire (see p. 152). The danger was not limited to urban expansion and green-belt out-of-town shopping centres (see p. 196). In 1973 Britain joined the European Union. This brought the country in line with Europe's Common Agricultural Policy and European farming subsidies, which gave financial incentives for farmers to produce more food. With entry into Europe came maximised farming production, intensification of mechanical farming methods, the introduction of pesticides known to damage natural habitats, and the removal of hedgerows – all altering the face of the 'cherished land'.

This was the tillable ground upon which the seeds of Euroscepticism were sown. Worse was to come. In 1975, the government published its Industrial Expansion Policy, which included the publication and free issue of a 'Government Aid for Industry' development map. This map identified vast swathes of Great Britain into which the government was prepared to financially support industrial expansion. In addition to the north-west, south Wales and the declining shipbuilding areas of Glasgow, the north-east and Merseyside, these areas included all of Wales, Scotland, Cornwall, everything north of Nottingham – that very same 'cherished land' exalted in 1973. Insets demonstrating the speed of freight implied that the remoteness of Northumbria and the Scottish Highlands had ceased to be their protection.

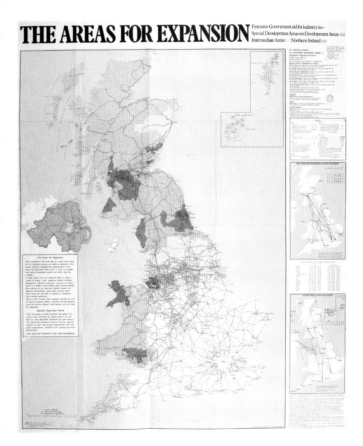

1974: 'From Liverpool to the World': the birth of Beatles tourism

THIS IS A modern pilgrim's map inspired by one of the most influential bands of the century: a guide to all sites of Beatles interest in their home city, Liverpool. It is both an official map and a retrospective one, commissioned in 1974 – four years after the band had broken up – by the City of Liverpool Public Relations Office.[110] It represents an early and far-sighted official appreciation of how Liverpool's vibrant music scene might help to revive the city's flagging fortunes, but it is also an appreciation of the recent past. Although ex-Beatles collaborated in various combinations on different projects, they would never perform together again. By 1974 they had visited India and grown their hair, but here the band is locked in 1968 – in the style of the animated film *Yellow Submarine* directed by George Dunning. Old Fred makes a guest appearance, hand in hand with the Chief Blue Meanie, and the Yellow Submarine itself bobs on the Mersey.

Beatles tourism is now a multi-million-pound industry. In the past year the gates of Strawberry Field, the former Salvation Army children's home, have been replaced by precise replicas to save them from further deterioration, and Ringo Starr's birthplace at 9 Madryn Street has been saved for the nation, although the surrounding terraced streets which place it in context are still scheduled for demolition. Both sites are featured prominently on the map.

In the early 1970s the situation was very different. The driving force behind the creation of the map was Ron Jones, then Deputy PR Officer for Liverpool City Council. He recalls that:

> it was impossible for a visitor to Liverpool to buy a Beatles souvenir or even a postcard at that time. There was no Beatles map or guide to Liverpool to tell visitors where they could find their homes, schools, the clubs and dance halls where they played or the places which inspired their songs. It would also be fair to say that at that time Liverpool did not have a reputation as a tourist city. The city's image was at its nadir – a grim northern city in decline, utterly without glamour, known for its slums, high levels of crime and unemployment and striking dockers and car factory workers.[111]

Jones describes himself as 'a "first-generation" fan of The Beatles', with the 1963 Cavern Club card to prove it, but his 'prime motivation

was not to glorify The Beatles but to promote Liverpool'. The map was one component in a comprehensive souvenir pack called 'The Beatles Collection – from Liverpool to the World', nine items which included posters, postcards, a discography and a short history of the band written by Jones' collaborator, London-based musician and music journalist Mike Evans.

Jones felt that he had to tread warily:

> at that time I think it would also be fair to say that Liverpool City Council, if not anti-Beatles, had no particular enthusiasm to honour them or to promote Liverpool on their backs. Aware of antipathy, even downright hostility from certain councillors, I decided not to put my plans before any council committee but to implement it and hope for the best!

In the event, the souvenir was well received and Jones was able to build on its success, eventually developing a Beatles 'infrastructure' of trained guides who gave conducted walking, coach and car tours for the increasing number of Beatle tourists who began to visit Liverpool.

The map was designed by Liverpool-based commercial art studio McCaffrey and Sharp. Established in 1963 by Stan McCaffrey and Jim Sharp, the firm counted Liverpool Council among its best customers; having worked with them before Jones was convinced that 'they were the right people to commission to do the artwork'. McCaffrey recalls that the 'Yellow Submarine' era was chosen in part because the film, with action divided between Pepperland and Liverpool, provided another direct link with the city. The Beatles do not seem to have been consulted individually, although their record company EMI was very helpful, but Jones assumes that they would have been sent copies out of courtesy. It was the first 'official' Beatles souvenir item specifically produced for Liverpool.

The Beatles map is not exclusively a celebration of The Beatles. Space has been found for the Diddymen and Jam Butty Mines of Knotty Ash, immortalised by Liverpool comedians Arthur Askey and Ken Dodd. Part of local lore, more recently they feature on Stephen Walter's 2008–9 map of the city.

1975: Nuclear power and the gamut of public opinion

ACCORDING TO THIS map, published by its international trade magazine, the nuclear power industry was thriving in 1975. And so globally it was, with as many power stations planned or under construction as were operational. Close-ups of North America, Europe and Japan detail the areas of their greater concentration.

It is not hard to see how in the mid-1970s nuclear power for civilian use was an attractive proposition for governments of developed nations. A world energy crisis had followed the oil crisis of 1973, the economic depression providing impetus for governments to seek alternative power sources. In Britain the issue was acute: industrial disputes and worries over the environmental impact of coal burning, together with the slow yield of North Sea oil (see p. 170), placed nuclear power at the top of the agenda.

But despite presenting a world vision of nuclear power through which the trade magazine hoped to demonstrate global enthusiasm, nuclear power would continue to suffer from an image problem. Its development during the Second World War by the United States military – the Manhattan Project – had produced the bombs dropped upon the Japanese cities of Hiroshima and Nagasaki in 1945 with enormous loss of life. From the outset, harnessing nuclear fission for civilian use was associated with danger, which a sequence of subsequent nuclear facility safety incidents would do nothing to reverse.

Benefiting from American cooperation, Britain's first nuclear power station, Calder Hall, had been opened in the north-west by the Queen in 1956. The following year the first disaster occurred, a fire at the neighbouring Windscale nuclear plant, which was not reported in the press. A partial nuclear meltdown at Three Mile Island in Pennsylvania, USA, in 1979 was followed in April 1986 by the catastrophic explosion and subsequent meltdown at the Chernobyl power plant in Ukraine.[112] The Chernobyl disaster affected a large portion of Europe, with radioactive waste carried through the air as far as Norway and the UK. The local population was contaminated by drinking milk from affected animals; a full appreciation of the spread of cancer and other associated cases has still to be established.

Following Chernobyl it would be far more difficult to convince a sceptical public of nuclear safety. On this map Chernobyl is just one of a number of large Western European nuclear facilities under construction positioned inland, near large sources of the fresh water needed for cooling. In general, however, the geography of nuclear power shows a preponderance of power stations on or near coasts. Some coasts, particularly that of California, with its fault line running underneath, were quite rightly agreed to be unsuitable.

So much for precautions. For many the danger lay in the inherent volatility of nuclear material, but the issue was also environmental and political, in a climate of international Cold War suspicion. In 1975 the most serious nuclear disasters were yet to occur. But even by then in Britain, enthusiasm for a third generation of power stations was stalling due to political delays, spiralling construction and safety costs, and the existence of remarkably effective popular opposition. According to a 1977 work published by the Institute of Nuclear Engineers, public opposition to nuclear power had been aroused by a misleading BBC documentary of 1971.[113] But opposition was spawned by more

than a television programme. This was the early 1970s, a period of popular environmental refocusing following the cultural revolution of the late 1960s and the profound effectiveness of photographs of Earth taken from space (see p. 154). The images of a supposedly fragile planet became a symbol and a spur for its protection.

Campaigns by protest groups Friends of the Earth, Greenpeace

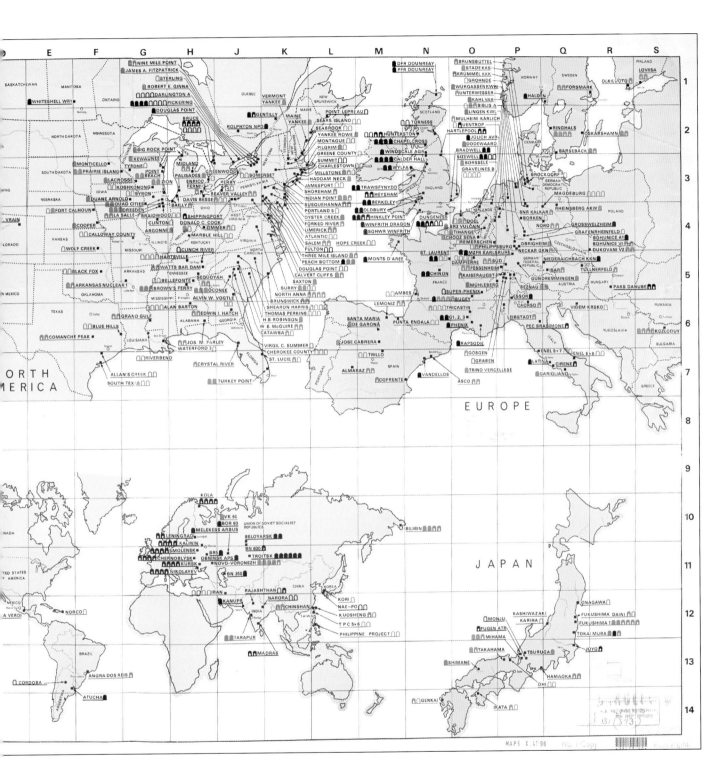

and the Campaign for Nuclear Disarmament (CND) (see p. 190) mixed articulate membership and popular lobbying with direct action against nuclear power and weaponry. The value of nuclear power as an environmentally sound alternative to fossil fuels was outweighed by its potent danger to life, its military associations and its inception at the point of nuclear holocaust.

1975: The North Sea oil bonanza

AS WITH THEIR treatment of terra firma, maps have imposed a variety of divisions upon the sea. On this Norwegian map, the North Sea has been divided and then apportioned into segments, which have been given further coloured subdivisions. These are licence areas for oil and gas fields that had been first discovered under the North Sea in the 1950s. The map was published by the American-owned Esso, which was then in partnership with the national Norwegian oil company Statoil. A general summary of theirs and their rivals' interests, with ESSO holdings coloured red, it served a variety of audiences. This copy, for example, was accessioned by the UK Ministry of Defence. It paints a geographical picture of intense regulatory governmental control over lucrative commercial and national interests, during a period of economic hardship.

Ratified by agreements in 1965, the continental shelf under the North Sea was divided up between its coastal countries, their sovereign territory extended beyond their coasts to artificial lines imposed on to the seabed. Those bearing the biggest smiles were Britain and Norway, those with lesser grins France, the Low Countries, Germany and Denmark. Without doubt least pleased was Scotland, or at least advocates of Scottish independence (the Queen's turning on of the first pipeline in Aberdeen in 1975 was Scottish police's biggest ever operation).[114] Initially, the shallower and calmer southern waters claimed by Holland were the most desirable as it was here that the first gasfields had been discovered. However, in 1969 the first giant oilfield, Ekofisk, was discovered by Philips Petroleum in the Norwegian sector.

Ekofisk is identifiable on the map as the intense Tetris-like collection of colours in the lower centre. By 1975, as we can see, Ekofisk was being well exploited by a variety of national and foreign companies in which the Norwegian government enjoyed a more-or-less 50 per cent stake. In 1976, it was producing 280,000 barrels of oil a day. But in Britain, once Ekofisk became successful, it became controversial. It was argued that the marking of Norway's portion of the continental

shelf had ignored the 'Norwegian trench', a 40-kilometre-wide passage of deep water (on the map it is a dark blue) which snaked around the bottom of Norway, cutting the shelf in two,[115] apparently limiting Norway's territory to the narrow strip within it.

If this was truly believed, it is easy to understand the statement that the division was 'perhaps the most generous present in English history'.[116] It would have been particularly galling because Ekofisk oil had to be piped first to Teeside in England, since the trench made laying a pipeline to Norway impossible. It wasn't just the oil which came over: young Norwegians came to Britain to study engineering in UK technical colleges. It was labelled by some in the north-east of England the 'second Viking invasion'.

Sovereign rights to North Sea oil were just the beginning. Beyond this primary division, a grid system divided the shelf into blocks to be licensed to oil and gas companies. Unlike the vast concession areas of Middle Eastern oilfields, these comparatively tiny licence areas reflected more stringent economic conditions and greater desire for tight governmental control, especially Britain's. Each area was roughly 100 square miles, but whereas Norway's concession blocks measured 15 minutes of latitude by 20 minutes of longitude, Britain's were smaller at 10 minutes by 12 minutes. Small areas limited the amount of success any oil company could have without needing to licence adjacent areas, which would be priced accordingly. In all cases, 50 per cent of licences reverted to the state after six years.

The grid pattern reflects Britain's desire to extract as much from the North Sea as possible, certainly compared with Norway at that time. Given the very different demands of these countries, their economies and populations, this is hardly surprising.[117] Yet, despite the North Sea oil bonanza, the revenue generated did not filter down to British society: 'Britoil' was privatised in 1982, and the resource was over-exploited. By attempting to describe the orderliness of North Sea oil, the map transgressed from reality. Natural carbon deposits do not develop in an orderly grid pattern, and the calm blue was nothing like the unpredictable, rough and dangerous North Sea.

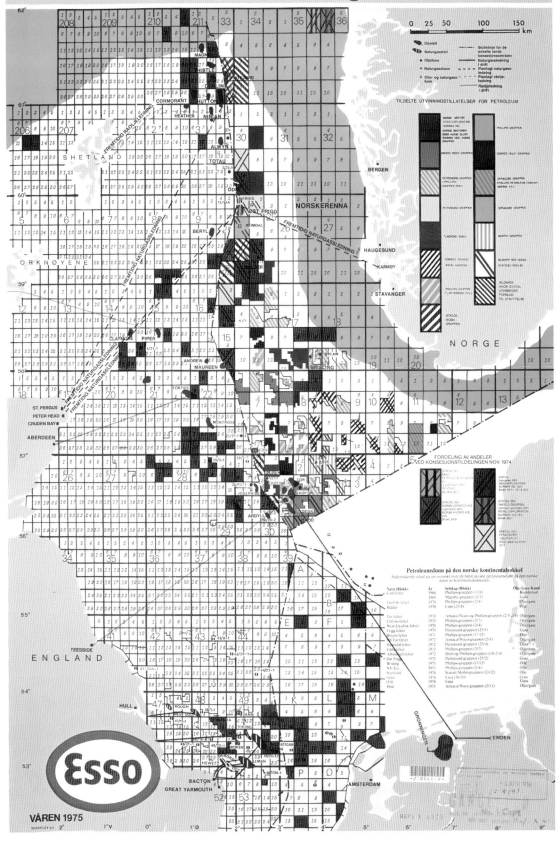

NORDSJØOLJEN
leteaktiviteter og funn

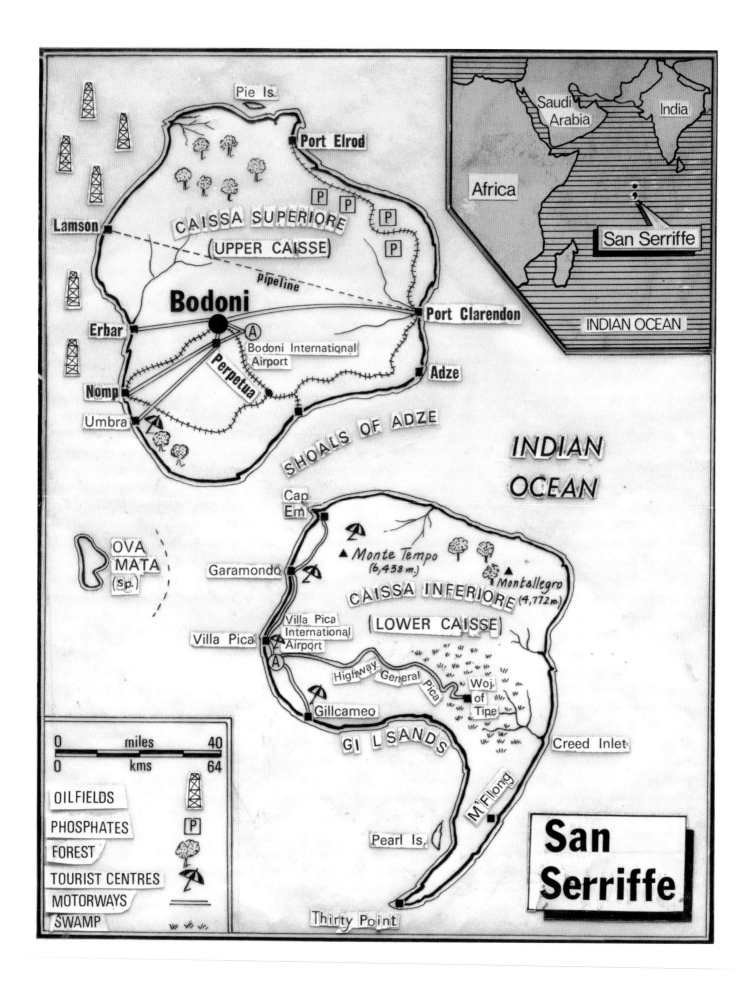

Pie Is.

Port Elrod

Lamson

CAISSA SUPERIORE

(UPPER CAISSE)

pipeline

Bodoni

Erbar

Port Clarendon

Perpetua

Bodoni International
Airport

Nomp

Adze

Umbra

SHOALS OF ADZE

INDIAN

OCEAN

Cap
Em

OVA
MATA
(Sp.)

Monte Tempo
(6,438 m.)

Montallegro
(4,772 m.)

Garamondo

CAISSA INFERIORE

(LOWER CAISSE)

Villa Pica
International
Airport

Villa Pica

Highway   General Pica

Woj
of
Tipe

Gillcameo

Creed Inlet

GILSANDS

M'Filong

Pearl Is.

| 0 | miles | 40 |
| 0 | kms | 64 |

OILFIELDS

PHOSPHATES

FOREST

TOURIST CENTRES

MOTORWAYS

SWAMP

Thirty Point

Saudi
Arabia

India

Africa

San Serriffe

INDIAN OCEAN

San
Serriffe

1977: April Fool's Day and the newspaper hoax: the islands of San Serriffe

AMONG THE MAPS of mythical places in the British Library is this original hand-drawn map of the islands of San Serriffe, published in the *Guardian* newspaper on 1 April 1977.

So wildly successful was the hoax in convincing people of the authenticity of this group of islands in the Indian Ocean that it has been credited with beginning the trend of humorous fake newspaper stories. April Fool's Day has a long history, and some equally preposterous pranks attached: the BBC's 1958 'spaghetti harvest' documentary, or the 1985 *Sports Illustrated* article about Sidd Finch, a baseball novice taught to pitch at 168 mph by a Tibetan monk. However, what marked San Serriffe out is that, once created, it became for all intents and purposes a real place, not only revisited in articles, described on enthusiast websites and even games, but integrated into the collective memory through conversations and reminiscences.

Thought up by the journalist Philip Davies, San Serriffe was elaborately described by a number of contributors to a seven-page *Guardian* supplement. As well as the map by Geoffrey Taylor, which included a key and an inset of the islands' location in the north-western Indian Ocean, a series of articles described its culture, economy and geography. Their president, for example, Maria-Jesu Pica, had come to power through a bloodless coup in 1971. Advertisements from brands such as Kodak and Guinness, very much in on the joke, added further authenticity.

Authenticity, and being placed off-guard by the unexpected humour of such ostensibly serious publications, may in part explain the repeated success of April Fool's Day hoaxes played on exactly the same day of the year. On reflection, the hoax was patently ludicrous. The Texaco advert included in the San Serriffe supplement, for example, offered a holiday to the islands' famous Cocobanana Beach. An article on the islands' geography explained that because of complex sand erosion moving the islands steadily eastwards, they were due to collide with Sri Lanka in 2011.

A second look at the map shows two islands bearing a remarkable similarity to a semi-colon. 'Sans serif' is, of course, a style of typeface. The typographical theme continues in the names of the islands (Upper Caisse and Lower Caisse) and in the articles where, for example, we learn of the national bird, the Kwote.

Revealing the hand of the newspaper in this way was part of the joke, but it revealed a great many more things besides the influence of absurdist comedy developed by such acts as Monty Python. It did not necessarily imply a lack of geographical knowledge for those of the newspaper's educated, left-wing readership base who failed to recognise the falsity of the islands. Perhaps still fewer would have heard of the genuine nearby UK-owned island of Diego Garcia, which had been forcibly depopulated in 1971 in order to be given over as an American military base.

The islands appeared at the right time, in the right place, and reflected particular post-colonial Britain's perceptions of what a small faraway island might be like. And very appropriately, in the age of factual travel programmes and magazine supplements, people contacted the *Guardian*'s offices on 1 April with holiday booking enquiries.

1977: The Silver Jubilee beacons: evoking the spirit of 1588

QUEEN ELIZABETH II's Silver Jubilee was celebrated with a series of extravagant events. A Silver Jubilee walkway and a fleet of silver buses materialised in London, and the Queen embarked on a marathon Commonwealth tour which surpassed in its number of destinations even her coronation tour of 1953 (see p. 128). For the main summer events, a specially appointed Queen's Committee re-cycled the 'New Elizabeth' theme of twenty-five years earlier with even more flamboyance. The scene was yet again Tudor England, the re-enactment of a 'royal procession' up the River Thames recalling Elizabeth I's regular journey between the Tower and the Palace, and a processional lighting of hilltop beacons on 6 June referencing the south coast warning fires that had announced the sighting of the Spanish Armada in the English Channel in 1588.

It was certainly an ambitious and extraordinary event, planned meticulously by the Royal Institute of Chartered Surveyors (RICS) alongside the military, the Ordnance Survey and the fire service. It began at 10 p.m. with the lighting of the Windsor Great Park beacon in the presence of the Queen, the royal family and a live television audience. Successive hilltop bonfires were then lit upon sight of the last, spreading down to Cornwall, up and around Wales and to Northern Ireland, through Scotland, even to St Kilda; the last was lit on the Shetland Islands one hour after the first. The guide contained photographs of bonfires and large crowds, at least in the more hospitable places. There was, apparently, 'much traffic congestion', a sure sign of success (see the 1927 solar eclipse, p. 72).

The obligatory map souvenir was published by Fairey Surveys. It showed the British Isles with the interconnected beacons and a great deal of embellishment, tapping effectively into the iconography of the wider celebration. An extract was included from Macaulay's nineteenth-century poem 'The Armada', and illustrations of 'scenes of invaders who helped to mould the present-day society of Britain'. In this it was not referencing multicultural Britain of the late 1970s. The map was not only a memento for participants, but also offered them the chance to be even more involved through its purchase: £6,000 from the proceeds of the sales went into a Silver Jubilee fund, which paid for the celebrations previously outlined. A special version of the map, 'mounted on rollers embellished with silver crowns', was presented to the Queen. But even the standard issue was regal: royal blue, with gold, ornate borders, and portraits of the Queen and Prince Philip. This was the sort of product one would expect to find for sale on the back pages of a Sunday tabloid colour supplement.

Not everybody, however, partook in the celebration. The 1977 backdrop to the jubilee was economic depression and rising inflation, labour disputes and sectarian trouble in Northern Ireland. Punk music group the Sex Pistols' vitriolic song 'God Save the Queen' articulated the anger, and their own mock boat party was broken up by police. Yet a backdrop of socio-economic malaise was precisely the sort of thing that royal, national celebrations were made for, to lift the national mood. They were recognised as an effective tool, funded by the state. The difference between 1977 and 1953 was that things did not look like they were about to get better.[118]

1952 · SILVER JUBILEE BEACONS · 1977

Buckingham Palace   The Queen's personal official residence

The Royal Standard

The Andrew's Cross of Scotland

Balmoral Castle
A Royal residence on Deeside in the highlands of Scotland

Palace of Holyroodhouse
Edinburgh's historic royal residence

A Viking Longship
Vikings, Angles, Saxons and Danes were early settlers in parts of Britain.

The Flag of Northern Ireland

Proceeds from the sale of this map will be donated to the Queen's Silver Jubilee Appeal. This is raising funds to support enterprises that give young people encouragement and opportunity to help others of all ages in the community.

The Triskelis
The emblem of the Isle of Man

The St. George's Cross of England

The Union Flag of Great Britain and Northern Ireland

Sandringham House in Norfolk is a favourite residence of the Royal Family.

The traditional emblem of Wales

WINDSOR

Windsor Castle
The oldest official royal residence

THE ARMADA
(An extract)

The forenoon breeze of eve exhaled that Sunset's amber tinct,
The purpling gauze of noontime rowed that laughter samite of gold.
Night sank upon the dusky beach, and on the purple sea.
Such night in England ne'er had been, ne'er again shall be
From Eddystone to Berwick bounds, from Lynn to Milford Bay,
The year of slumber was as bright and busy as the day.
For swift to east and swift to west the ghastly war-flame spread,
High on St. Michael's Mount it shone, it shone on Beachy Head.
Far on the deep the Spaniard saw, along each southern shore,
Cape beyond cape, in endless range, those twinkling points of fire.
The Eddar felt his skiff or ere our Tamar's glittering waves
The rugged miners poured to war from Mendip's sunless caves.
O'er Longleat's towers, o'er Cranbourne's oaks, the fiery herald flew
He roused the shepherds of Stonehenge, the rangers of Beaulieu.
Right sharp and quick the bells all night rang out from Bristol town,
And ere the day three hundred horse had met on Clifton down;
The sentinel on Whitehall gate looked forth into the night,
And saw o'erhanging Richmond Hill the streak of blood-red light.
Then bugle's note and cannon's roar the deathlike silence broke,
And with one start, and with one cry, the royal city woke.
At once on all her stately gates arose the answering fires;
At once the wild alarum clashed from all her reeling spires;
From all the batteries of the Tower pealed loud the voice of fear;
And all the thousand masts of Thames sent back a louder cheer;
And from the furthest wards was heard the rush of hurrying feet,
And the broad streams of pikes and flags rushed down each roaring street:
And broader still became the blaze, and louder still the din,
As fast from every village round the horse came spurring in.
Till Belvoir's lordly terraces the sign to Lincoln sent,
And Lincoln sped the message o'er the wide vale of Trent;
Till Skiddaw saw the fire that burned on Gaunt's embattled pile,
And the red glare on Skiddaw roused the burghers of Carlisle.

1832   Macaulay

A Galleon of the size of the Spanish Armada. The Armada was routed by the English fleet in 1588. The alarums ashore are described in Lord Macaulay's poem.

The Norman Invasion was the last to succeed on English soil. This section of the Bayeux Tapestry portrays the Battle of Hastings in 1066.

The Roman Invasion

© The Royal Institution of Chartered Surveyors. Produced by Henry Stevens Ltd, Sherborne, Dorset.

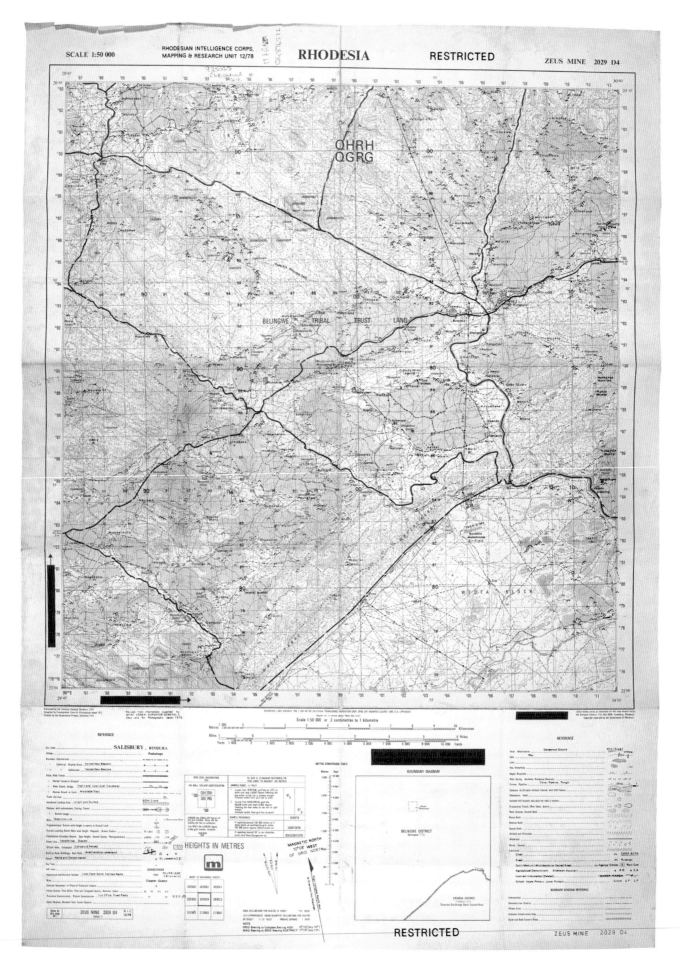

1978: Colonial mapping and the Rhodesian Bush War

THIS IS A large-scale (1:50,000) survey of a sparsely populated but mineral-rich corner of Rhodesia. By 1978 Rhodesia was Britain's last and most controversial formal colony in Africa, shortly to achieve full independence as Zimbabwe. Produced by the Rhodesian Intelligence Corps during the final, bloody escalation of the Bush War, and marked 'restricted', it was later used by an academic researcher into global change, who was comparing it with satellite imagery as part of the Landsat programme.

The caveat 'users please note that some of the overprinted information is unchecked' is reminiscent of the warning on the Boer War map of Bloemfontein (p. 18), but the quality of the cartography is a world away. Layers of detailed information are supplied about airstrips, the different kinds of roads and crossings (from tarred roads to 'motorable' tracks), different densities of bush, the location of settlements (from cities to kraals), and religious establishments (from churches to 'witch doctors', 'spirit mediums' or 'sacred areas'). Despite the regionally specific data, it remains peculiarly British in its style.

Land surveys in Zimbabwe go back to the early 1890s, recording the claims of the earliest European arrivals. The dismal showing of maps issued in the Boer War led to renewed, if fragmentary, efforts by the various colonial survey agencies, but real change came only after 1945 when a centralised body, the Directorate of Colonial Surveys (DCS), was created.[119] The DCS emerged from wartime rivalry between the Geographical Section General Staff and the Ordnance Survey. Just as it became clear to the rest of the world that a bankrupt and starving Europe could no longer expect to hang on to its colonies, Britain finally accepted that consistent, detailed mapping was vital for the long-term development of all its overseas territories. Only an estimated 400,000 of some 2,250,000 square miles had been systematically mapped and, in close cooperation with the RAF, the DCS undertook an ambitious campaign of mapping to fill in the gaps, based on fieldwork and modern aerial surveys. In the twenty years after 1945 almost all of that vast territory was transformed from Empire to Commonwealth. The DCS became the DOS, Directorate of Overseas Surveys, but the commitment to supporting the development of former colonies remained.

The mapping of Rhodesia in 561 sheets on a scale of 1:50,000 was complete by 1970, on the transverse Mercator projection (ideal for the highly accurate mapping of relatively small areas).[120] Mineral wealth lured the first European settlers to the region and the focus of this sheet is the Zeus emerald mine, named by someone classically minded, and later known as Sandawana. Opened in the 1950s, it closed in 2012.[121]

Rhodesia was initially controlled by private enterprise, the British South Africa Company (inspired by the former East India Company). Even after Rhodesia became a Crown Colony in 1923, the settlers retained an extraordinary degree of autonomy, and their repressive treatment of the indigenous population was always controversial. One by one neighbouring African colonies gained their independence, but granting independence to Rhodesia without black majority rule was politically unacceptable in the UK (although public opinion was, as ever, divided and there was considerable sympathy for the settlers). In 1965 the Rhodesian government, led by wartime fighter pilot Ian Smith, took matters out of the British government's hands, sure that British military intervention was unthinkable. The Unilateral Declaration of Independence was the first of its kind since 1776, although the Rhodesian document concluded with 'God Save the Queen' and its purpose was to perpetuate white minority rule – albeit with the support of a section of the black population which also feared the consequences of majority rule.

Smith's illegal regime flourished for over a decade – a product of Cold War politics as well as support from apartheid-era South Africa and Portuguese Mozambique. Black nationalist guerrilla forces were trained and supplied by the Soviet bloc, but UN sanctions were routinely flouted by the West, including the US, which needed Rhodesian chrome and nickel in Vietnam. However, by the end of the 1970s international support had drained away, the guerrilla forces were more numerous and better armed than ever, and Margaret Thatcher's new Conservative government was able to broker a negotiated settlement.

Created rather late in the day, but with the best of intentions, our detailed survey was pressed into service in a vicious underground war. It is as good a reminder as any in this book of the incoherent and piecemeal way in which the British Empire was acquired, administered and finally dismembered.

1980: Happy Eater: catering for Britain's A roads

HAPPY EATER WAS a chain of inexpensive road-side restaurants established in 1973 in competition with Little Chef. As this route planner shows, Happy Eaters were chasing the same market: the restaurants were all sited on A roads, which were still commonly used for long-distance travel in preference to motorways. The menus were hardly worlds apart either, being variations on all-day English break-fasts, and fish and chips. Writer Joe Moran deplores the quality, quoting the 'probably apocryphal story of the customer who ordered an omelette and was told he couldn't have one "because they haven't sent us any"'.[122] The brochure stresses reasonable prices, a clean environment and car parking.

Happy Eaters were marketed as family restaurants, outmanoeuvring Little Chef by making a point of catering for fractious children on long car journeys. Much of the supporting artwork on the map stresses this. Outdoor playground equipment – the giant elephant slide next to the restaurant – is visible on the cover, and both children in the cheerful nuclear family in the foreground are sporting Happy Eater merchandise in the form of t-shirts and pin badges. On an inside panel, more contented children, clearly sampling the special children's menu, have been issued with colouring games and solitaire. It may be that the logo, reminiscent of Pac-Man, was also intended to be child-friendly, but it was also the inspiration for a piece by artist Tim Head described as a 'consumerist nightmare of cannibalistic proportions'.[123]

In 1981, Happy Eater was purchased by the hotel division of Imperial Group Plc – part of the ongoing diversification of Imperial Tobacco, as a response to declining demand for tobacco products.

Twenty-one Happy Eaters existed at the time of the deal, and twenty-one are marked on the map. One panel also advertises Pickard Motor Hotels, a new venture by Happy Eater founder Michael Pickard, so it is reasonable to assume that our map reflects the state of play just before the deal.

The chain was taken over by its great rival Little Chef in 1986 and finally disappeared from the roadside in 1997, predeceasing the premiership of its most famous patron, Prime Minister John Major, by a matter of months. In 1991, Major stopped for a fry-up at the Happy Eater on the A1 in Doncaster on his way to the Young Conservative Conference in Scarborough. Proposals to deregulate the building of motorway services as part of his Citizen's Charter later that year were inevitably dubbed 'the Happy Eater Charter'.

Politicians seeking to establish working-class credentials and families were not the only clientele. Musician John Barrow describes motorway service stations as 'an absolute prerequisite for life on the road'; even though they were expensive compared with 'greasy joe' transport cafes: 'you needed a mortgage for an extra sausage!' Despite which, 'many an after gig post mortem was conducted in a Little Chef or Happy Eater in the twilight hours. I often contented myself with a strong mug of coffee but these pit stops were essential, driving monotonous motorway miles was potentially dangerous for a tired driver'.[124]

It is not only the Happy Eaters which have subsequently vanished. Paper maps like this one are far less common today, although branch or restaurant locators are now an integral part of most retail websites. The need for a map is as great as ever, but the delivery could hardly be more different.

Happy Eater FAMILY RESTAURANTS ROUTE PLANNER

We're happy to serve you!

HAPPY EATER FAMILY RESTAURANTS

KENT
1	A20	CHARING North of Ashford
2	A299	WHITSTABLE Thanet Way

SURREY
3	A25	BETCHWORTH East of Dorking
4	A217	BURGH HEATH South of Sutton
5	A3	HINDHEAD South of Guildford
6	A31	HOGS BACK Seale East of Farnham
7	A24	HOLMWOOD South of Dorking
8	A23/M23	HOOLEY North end of M23
	A3	KINGSTON NORTH Kingston-By-Pass London End
9	A3	KINGSTON SOUTH Kingston-By-Pass London End
10	A247/A3	RIPLEY North of Guildford

SUSSEX
11	A22	FELBRIDGE North of East Grinstead
12	A23	HANDCROSS South end of M23
13	A21	LAMBERHURST South of Tunbridge Wells

HAMPSHIRE
14	A3	RAKE North of Petersfield

SOMERSET
15	A303	CAMEL CROSS West of Wincanton
16	A30	HENSTRIDGE East of Sherborne

DEVON
17	A38	KENNFORD NORTH West of Exeter
	A38	KENNFORD SOUTH West of Exeter

OXFORDSHIRE
18	A34/A44	CHIPPING NORTON North of Oxford

HERTFORDSHIRE
19	A10	HIGH CROSS North of Ware/Opening September 81

CAMBRIDGESHIRE
20	A1	PETERBOROUGH Norman Cross near Peterborough/Opening July 81

YORKSHIRE
21	A64	YORK South West of York/Opening August 81

A1 TUXFORD

KEY
———	Motorways
———	Primary Routes
———	A Roads
○	Primary Towns

ENGLISH CHANNEL

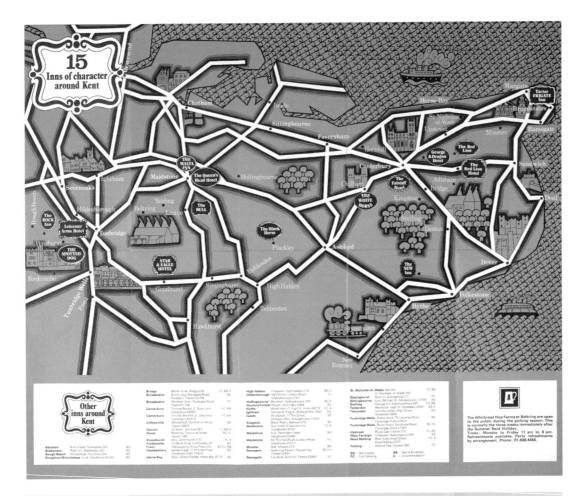

15 Inns of character around Kent

Other inns around Kent

The Whitbread Hop Farms at Beltring are open to the public during the picking season. This is normally the three weeks immediately after the Summer Bank Holiday.
Times : Monday to Friday 11 am to 6 pm. Refreshments available. Party refreshments by arrangement. Phone: 01-606 4455.

BS – Bar Snacks
FC – Full Catering
BB – Bed to Breakfast
A – Accommodation

Tartar FRIGATE Inn

The Red Lion

THE MALTA INN

The Black Horse

THE WHITE HORSE

The Falstaff Hotel

The NEW INN

The Red Lion Hotel

The Queen's Head Hotel

STAR & EAGLE HOTEL

THE SPOTTED DOG

George & Dragon Hotel

THE ROCK INN

Leicester Arms Hotel

The BULL

1981: Drinking and driving: fifteen inns of character in Kent

THE QUINTESSENTIAL OAK-BEAMED English inn, set in historical south-east England, was to get a makeover in this incisive and dynamic advertisement map by the Whitbread Brewery Company. A piece of ephemera, given away at all of the fifteen Whitbread-owned pubs marked on the map, it would have been picked up by tourist-travellers passing through the region en route to France via ferry, or on day trips to historic Canterbury and the differing seaside experiences of Pegwell Bay and Margate. The map is drawn in a cutting-edge style, while tradition is evoked by the appeal of the country inn.

We have seen these straight, schematic transport lines before in the 1963 railway poster for the Yorkshire coast (p. 121). By 1981, it was roads and motorcars that were taking families on their visits to the seaside, cathedrals, castles and oast houses. These latter are the highly appropriate (for breweries) buildings with the tall angled chimneys. Whitbread aimed to attract these daytrippers to their inns. They do this here by suggesting that it is these inns of character – their variety displayed in their symbols (notably absent from the modern incarnation of 'chain' pubs) – that are the main Kentish attractions.

In a real sense the map is correct. Inns remain important historical sites both for travellers and in communities, and many are centuries old. The Falstaff Hotel in Canterbury that is illustrated here, for example, was built in 1403. Both the building and its cultural significance were reason enough to visit, while for Whitbread there was value in owning and promoting such a 'flagship' inn in terms of profit and image. Whitbread themselves had a long pedigree, having been founded in 1742. Already one of the biggest breweries in the UK, from the 1970s their investment and diversification included purchasing a stake in the Southern Television Company. However, they were also criticised for buying up smaller, independent breweries.

An advert portraying interesting inns serving local brew offset the image of the large corporate company. Any disingenuousness probably wouldn't have registered. The unlikely coexistence of a hovercraft in the Channel and the steam engine on the coast perfectly demonstrates how well contradictory messages can work in advertising. Strong advertising was a key feature of the alcohol industry, immensely important to alcohol brands such as Guinness, Jameson and Babycham. Indeed, corporate advertising, exemplified by the huge success of the company Saatchi & Saatchi, became one of the hallmarks of Conservative Britain of the 1980s. Traditional or modern, the consumer took from it what they wanted.

Seeking to reconcile such differences as old and new, local and national (not to mention drinking and driving) is missing the point of this map. It didn't require great capital to produce. It was cheap, ephemeral, contradictory and aimed to please everyone. It was an image, like a television snapshot, a magazine cover, a glimmer of the promise of leisure time, places of historical interest, a car journey, relaxation with a drink. This was what people wanted, and for landlords, breweries, parent companies and their advertising agencies it was very profitable.

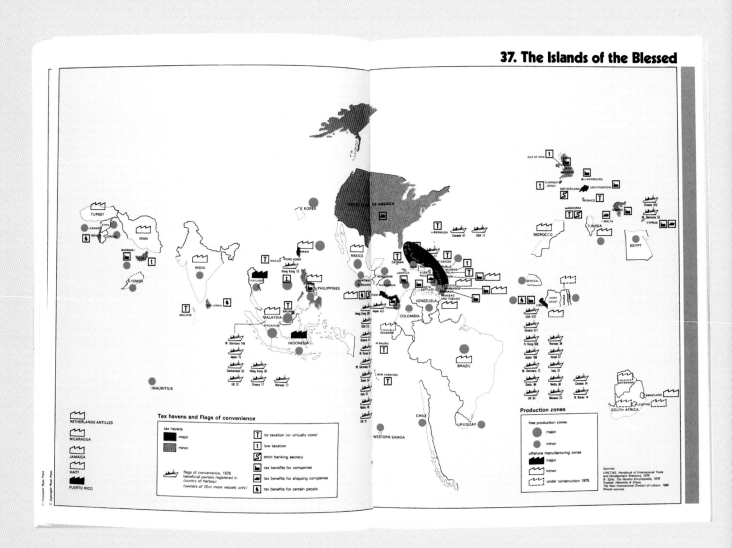

Tax havens and Flags of convenience

tax havens
- ▨ major
- ▨ minor

- T no taxation (or virtually none)
- T low taxation
- S strict banking secrecy
- 🏭 tax benefits for companies
- 🏭 tax benefits for shipping companies
- 🏭 tax benefits for certain people

flags of convenience, 1978
beneficial owners registered in
country of harbour

(owners of 10 or more vessels only)

Production zones

free production zones
- ● major
- ● minor

offshore manufacturing zones
- ▰ major
- 🏭 minor
- 🏭 under construction 1975

Sources:
UNCTAD, Handbook of International Trade
and Development Statistics, 1979
B. Spitz, Tax Havens Encyclopedia, 1978
Fröbel, Heinrichs & Kreye,
The New International Division of Labour, 1980
Private sources

NETHERLANDS ANTILLES

NICARAGUA

JAMAICA

HAITI

PUERTO RICO

1981: 'Islands of the Blessed': a socialist map of tax havens

SLANDS OF THE Blessed' is one of a number of political map statements in the provocative *State of the World Atlas* of 1981. Published by the left-leaning Pluto Press in the wake of the late-1970s economic downturn, the atlas espoused a socialist world view through a series of thematic maps illustrating imbalances caused by Western militarism and capitalism. These included the possession of the world's wealth in the hands of the minority, pollution, exploitation, arms sales and inequality. In the words of the atlas's introduction, 'the destructive aspects of the state have come crucially to exceed the constructive ones'.

None of the maps pulls many punches, but 'Islands of the Blessed' is particularly effective in portraying the inequality inherent in a global economy for which geographical borders and distances had quite literally ceased to exist. Its focus is upon the tax-avoidance measures of big business through international tax havens, and free and offshore production zones.

Tax havens are places – usually small and quiet sovereign states – which through loopholes and incentives enable individuals and companies, with their assets and profits, to be registered outside their own countries. There are strong financial benefits, with little or no taxation, as well as the added bonus of secrecy. Their emergence as serious money-saving enterprises coincided with the period of economic prosperity from the 1950s and an evaporation of protectionist policies such as the gold standard, which had been felt so essential before 1932 (see p. 82). Over the 1960s, multinational corporations became adept at spreading and hiding their investments and profits behind complex legal structures, which became particularly useful during the economic crises of the 1970s.

Offshore and free production zones (dotted green) permitted trade and manufacture without the scrutiny and taxation of the state in which they were based; labour in these areas was often very cheap (and the populations some of the poorest on earth). The majority are in Central and South America, South East Asia, and coastal Africa. Eric Hobsbawm noted one such zone in the middle of the Amazon rainforest.[125]

This was the geography of capitalism, the network of global business for which distance or obstacle was no hindrance. Exporting goods on ships, registered in improbable places for tax reasons, was so cheap that even great distances did not affect profits. Unlike the great European empires of previous centuries, globalisation transcended national borders and made the spaces in between utterly irrelevant.

Pluto's maps were socialist statements, the proliferation of which coincided with British social and economic deprivation, particularly in inner cities and in areas of declining heavy industry. The strikes and protests of the 1978–9 'Winter of Discontent' saw Jim Callaghan's Labour government replaced by Margaret Thatcher's Conservatives. Labour experienced a swing to the left. Pluto Press, linked with the Socialist Workers' Party, was established in London in 1969 and published material throughout the 1970s and 1980s, including literature for trade unions and books such as *Farewell to the Working Class*. *The State of the World Atlas* was their most popular work.

Pluto and others were preaching to a converted who saw taxation as necessary to fulfil the modern state's obligation to provide for its people. Their ideology, which tends to rear up during times of hardship, viewed tax havens and big business as implicitly wrong and immoral. Yet the opposite ideology, which became prevalent in the 1980s climate of free-market capitalism, saw things differently. As an instruction manual on how to use tax havens published in 1978 stated, 'if you think that tax travellers should stay at home and buckle down to the job of paying taxes, pause to remember that they may be seeking to preserve their dignity as well as their fortune'.[126]

1982: FAGA and the Falklands

THIS IS THE second printing of a commemorative post-card by Frederick Arthur George Amos Foley (1911–89), known as FAGA. It is a striking contemporary British response to the Falklands War of 1982, showing the Prime Minister, Margaret Thatcher, watching over British possessions in the Antarctic. A couple of trivial factual errors were rectified for this version and only British territory thought to be claimed by Argentina was indicated with the customary red: the South Orkneys and South Shetlands were excluded on the basis of advice given to the artist that they were not claimed and therefore not the subject of the first paragraph of a speech by Thatcher quoted on the postcard, a 'warning to aggressors', delivered in the House of Commons on 15 June 1982.[127]

This uncertainty about faraway countries seems entirely in keeping with contemporary perceptions of the crisis. It was the war itself that put the Falklands on the map in Britain. The Defence Secretary at the time, John Nott, revealed in his autobiography that he struggled to find the Falklands on the globe in his office, and he was dismayed to discover just how far away they were.[128] Official papers, recently released under the thirty-year rule, reveal just how divided the government of the day was. The impending withdrawal of the ice patrol vessel HMS *Endurance*, depicted by FAGA in the foreground, was one of the immediate factors which mistakenly persuaded the Argentinians that the British would not take military action to recover the islands. However, they were not mistaken in their belief that over a much longer period elements of the British government were searching for a dignified way of sharing sovereignty or withdrawing altogether. Even after the invasion of the islands by the military dictatorship which then ruled Argentina, some British politicians and their advisors still advocated a compromise such as compensating the islanders and resettling them, fearing the greater humiliation of a second Suez.

Scrambling a viable task force within forty-eight hours (augmented as further vessels became available), sending it to the far side of the world and – crucially – winning the war was a remarkable gamble. The war of words which broke out between British tabloid newspapers the *Sun* and the *Daily Mirror* is just one of the more extreme indications that the press and public were no more united in their responses than the politicians. However, if Galtieri's Argentinian junta had been seeking a distraction from domestic politics, so had Thatcher. The British victory drew a line under a decade of defeat: strikes, inflation, and then the riots and recession that had characterised Thatcher's first years in office. Victory secured the future of the Thatcher government and made her a player on the world stage; it also gave her (and every succeeding British prime minister) confidence in military intervention overseas. Attitudes to the Falklands themselves have, if anything, hardened in the aftermath of the conflict. It would be career suicide today for a mainstream British politician to deny the islanders' right to self-determination.

In that context FAGA's postcard can certainly be said to reflect a majority view at the time it was created, but it is difficult to determine how wide its audience was. As artist, calligrapher and postcard designer, FAGA championed the 'moderns' over the established vintage 'classics' in the rarified but polarised world of British postcard collecting. After service in the Royal and Merchant navies, FAGA studied calligraphy at the City & Guilds College of Art in his native London. His trademark designs weave detailed, elaborately lettered commentary together with multiple images. Between the 1960s and 1980s he created around two hundred commemorative cards, some on historical themes but others, such as this one, recording contemporary events. The Falklands have always been popular with philatelists, and FAGA's series of six Falklands War cards would have appealed to collectors of both postcards and stamps, as examples exist with special commemorative handstamps (for example, the dates of the Victory Parade and, on our example, Mrs Thatcher's visit to the islands on 11 January 1983). However, they also seem to have captured the public mood, and according to FAGA's friend and executor Ron Griffiths, they were singled out for praise in the *Daily Telegraph* and the *Financial Times*.

"WE DO NOT NEED TO NEGOTIATE IN ANY WAY WITH THE UNITED NATIONS OR anyone else about the British sovereignty of these islands.

Our forces did not risk their lives for UN trusteeship. They risked their lives to defend BRITISH TERRITORY, the BRITISH WAY of LIFE, & the right of BRITISH people to determine their own future

I hope we have restored once again the dominance of BRITAIN, & let every nation know that where there is BRITISH SOVEREIGN territory it will be well & truly defended & will never again be the victim of aggression"

Mrs. M.Thatcher, addressing the House of Commons 15 June 1982

FALKLAND ISLANDS

SOUTH GEORGIA ① LEITH GRYTVIKEN
ARGENTINIAN INVADERS EVICTED 25 APRIL 1982

SOUTH SHETLAND ISLANDS

SOUTH ORKNEY ISLANDS ②

SOUTH SANDWICH ISLANDS

BRITISH ANTARCTIC TERRITORY

③
④
⑤
⑥

GRAHAM LAND

PALMER LAND

ANTARCTIC PENINSULAR

WEDDELL SEA

HALLEY BAY ⑦

CAPT. JAMES COOK·1728-79 in "RESOLUTION" 1774

H·M·S "ENDURANCE"
ICE PATROL SHIP FOR ANTARCTIC WATERS: ENGAGED IN HYDROGRAPHIC & OCEANOGRAPHIC SURVEYS & ACTING AS SUPPORT SHIP FOR BRITISH ANTARCTIC SURVEY & GUARD SHIP

TONNAGE 3,600
SPEED·14·KNOTS
2·HELICOPTERS
2·20mm GUNS

FAGA

A171

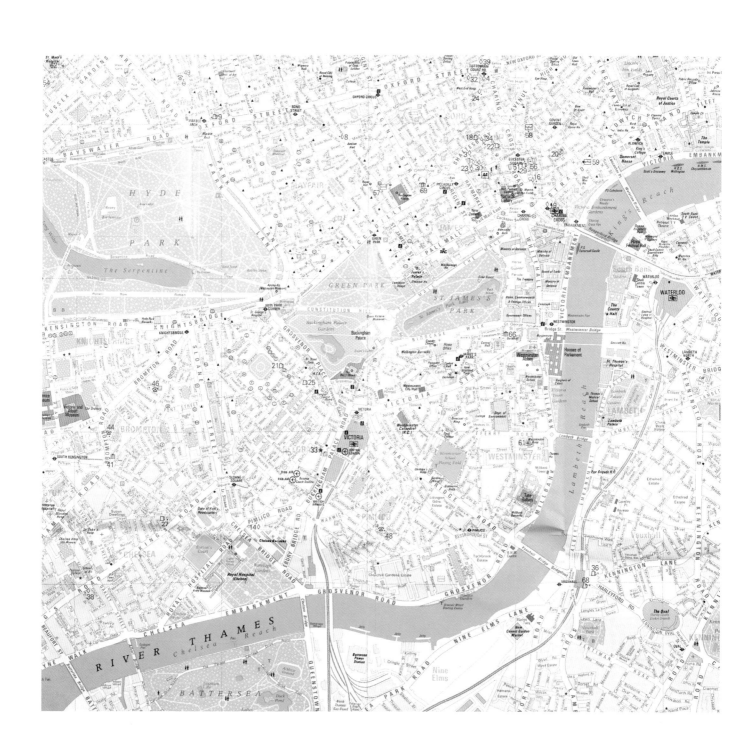

1982: Gay London

THE 'LONDON GAY City Map' was issued in a clear plastic wallet with the 1982 *Spartacus International Gay Guide for Gay Men*. *Spartacus* began life as a monthly gay magazine, published from a Brighton boarding house. The first of the annual guides appeared in 1970 and the publication moved to Amsterdam in 1973, possibly because prosecution there under obscenity laws was less likely. Since 1987 it has been published by Bruno Gmünder Verlag in Berlin.

The map is a 1978 John Bartholomew and Son map of London, overprinted in red with symbols denoting gay or sympathetic bars, clubs, restaurants, bookshops and other places of interest. The text states that it has been 'expertly prepared by Britain's leading map producers' and this emphasis is telling. Such maps as exist in the earlier editions of the guides were traced and copied without attribution. It is significant that a mainstream publisher was willing to be associated with it. Even if gay publications did not fall foul of the law, finding a major retailer or distributor had been difficult. W. H. Smith, for example, had initially refused to stock or distribute *Gay News* in 1972, but in the intervening decade the social landscape had begun to evolve.

The guide was 'written by Britain's leading expert David Seligman, who knows most of the places intimately'. Seligman was a pioneer of gay journalism, a founder of *Gay News* and the Gay Switchboard (an advice line). Guide and map cover far more than what Seligman terms 'the gay life', with information about shopping, theatres and museums – the things every tourist needs to know. The overprinted 'gay' venues on the map support Seligman's remarks that London was unusual, when compared with other European cities, in the immense variety of gay establishments and their general distribution throughout the city: although there were concentrations in the West End and Earls Court, there was 'no specific gay area'.

A slight difficulty presents itself in that the numbers on the map do not relate to anything in the guide, but it is still possible to use them effectively together. The map features pubs such as the Salisbury on St Martin's Lane, 'the celebrated gay, theatrical and show business pub. Always crowded and cruisy ...' (and with 'excellent but expensive' sandwiches). Heaven, on Villiers Street, 'one of Europe's largest and most elegant gay discos', is described as the 'in place' of 1980: 'the beauty of the waiters compensates for the long wait for your drink'. There seem to be very few omissions, but Adams in Leicester Square, which became Subway in 1981, is conspicuous by its absence. Possibly it was a little too wild, even for Seligman.

In the 1970s there had been tension between commercial and community venues, publications and events: some activists felt that no one should be profiting from the gay community. No distinctions of that nature seem to be made here. The range included is enormous and often highly specific, featuring venues with transvestite or other shows, 'leather' venues, places known for 'heavy military patronage' or where one might encounter 'Rough Types'. Venues which were 'gay- and lesbian-mixed' are differentiated from those where one or other were 'not welcome or excluded', as are venues which either welcomed or discriminated against older gay people. The gay life appears to have been flourishing, but there are occasional jolts for the modern reader. Hotels where two men may sleep together and discos where two men may dance together are singled out for special mention, and the guide uses the chilling abbreviation AYOR, 'At your own risk', to denote a 'Dangerous place with risk of personal attack or police activity'.

There is a terrible poignancy attached to the year of publication. The name AIDS was coined in 1982 for an incurable and (at the time) swiftly fatal disease, whose first victims were mostly intravenous drug users and, increasingly, gay men. The ensuing public health crisis created a climate of fear and a moral panic, which reached its nadir with the passing of Section 28 of the Local Government Act 1988, explicitly prohibiting the 'promotion' of homosexuality by local authorities in schools and elsewhere. The legislation was only fully repealed in 2003.

1984: A map of George Orwell's '1984'

WHEN GEORGE ORWELL'S futuristic novel *Nineteen Eighty-Four* was published in 1949, the world was still acclimatising to the fact that it had been irrevocably changed by six years of global conflict. Two power blocs with opposing ideologies of East and West, communism and capitalism, faced each other down in incidents such as the 1948 Berlin Airlift. The Korean War had followed the next year.

Orwell developed a geopolitical scenario as the backdrop to his dystopian future, in which three superpowers (the newly independent India and the populous 'yellow peril' of China constituted a third) were perpetually at war with each other so as to maintain industrial production and control over their populations. The novel's context of deprived freedoms against restrictions of the state, expressed through the struggle of the character Winston Smith, also referenced post-war Britain, where food was still rationed and major amenities remained under government control.

That was 1949. Yet such was the resonance of Orwell's work that the arrival of 1984 provoked a variety of re-engagements and evaluations. A film of the novel starring John Hurt was released, while music group Eurythmics' song 'Sex Crime Nineteen Eighty Four' offered a provocative take on casual sex, a punishable offence in Orwellian Britain. Although the new world order described by Orwell had not demonstrably come to pass, it was very much in David Llewellyn's mind when he drew his world map on Briesemeister's oblique projection (devised in 1953) and called it 'The World of George Orwell's "1984"'.

What does the map tell us about the coming of age of Orwell's vision? It certainly demonstrates that the novel's geopolitical scenario had resonance in the mid-1980s. It also weds literature with science, since it was drawn by Llewellyn as part of his geography and surveying course at Wembley Technical College. Not the dystopian concrete landscape of Wembley, Middlesex, but Wembley, Western Australia (its own desert 'dustopia' had been articulated by the *Mad Max* film franchise from 1979). It was one of five student maps published in *Cartography of Western Australia* and presented to the 12th International Cartographic Association (ICA) conference in Perth in 1984.[129] It filled the imaginary map category, the others being categories of topographical, urban, marine and statistical map. However 'imaginary', the quality of Llewellyn's map was very much in keeping with cartographic excellence promoted by the ICA and especially their incumbent president, the American geographer Arthur H. Robinson (1915–2004). Robinson had worked extensively on map projections and promoted technical standards in map-making with particular zeal, arguing against human interventions in what he saw as a scientifically pure process.[130] Llewellyn's map is the sort of exemplary student cartography that would have scored highly.

Ironically, scientific rigour was at variance with one of the central messages of *Nineteen Eighty-Four*: humanity, expressed through Winston Smith's quest for warmth and compassion. These were the themes that provided Orwell's novel, as well as other dystopian novels such as Aldous Huxley's *Brave New World* (published in 1932), with their enduring power. The world order of *Nineteen Eighty-Four*, firmly rooted in post-war 1949, had moved resolutely into fantasy.

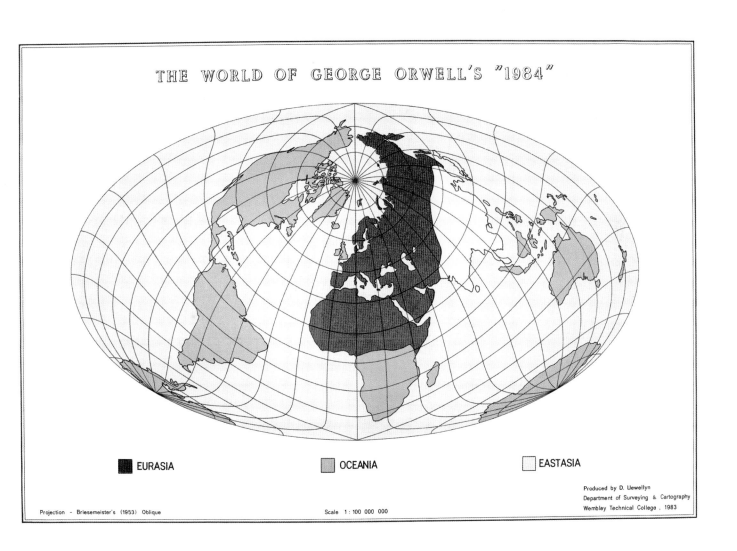

THE WORLD OF GEORGE ORWELL'S "1984"

■ EURASIA   ▨ OCEANIA   ☐ EASTASIA

Produced by D. Llewellyn
Department of Surveying & Cartography
Wembley Technical College , 1983

Projection - Briesemeister's (1953) Oblique   Scale 1 : 100 000 000

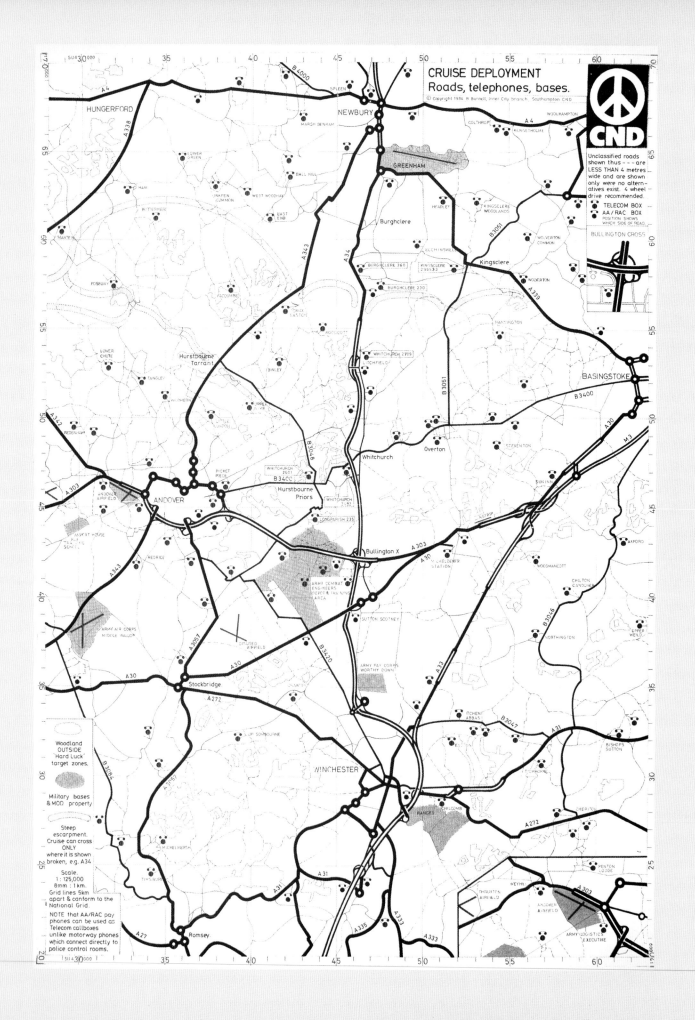

CRUISE DEPLOYMENT
Roads, telephones, bases.

© Copyright 1984 B. Burnell, Inner City branch, Southampton CND

Unclassified roads shown thus - - - are LESS THAN 4 metres wide and are shown only were no alternatives exist. 4 wheel drive recommended.

- TELECOM BOX
- AA/RAC BOX
  POSITION SHOWS WHICH SIDE OF ROAD.

BULLINGTON CROSS

Woodland OUTSIDE 'Hard Luck' target zones.

Military bases & MOD property

Steep escarpment. Cruise can cross ONLY where it is shown broken, e.g. A34

Scale.
1 : 125,000
8mm : 1km.
Grid lines 5km apart & conform to the National Grid.

NOTE that AA/RAC pay phones can be used as Telecom callboxes unlike motorway phones which connect directly to police control rooms.

1984: A CND operations map: guerilla warfare on Salisbury Plain

A GLIMPSE OF the organisational extent of 1980s anti-nuclear protest is provided by this specially produced tactical map. Printed by the Campaign for Nuclear Disarmament (CND) in 1984, it shows the area south of Greenham Common airbase in Berkshire, which since 1981 had been a site for the storage of American Cruise missiles. Key features are the airbases, in red, and the strong black lines of the roads along which lorry convoys – the target for protestors – would travel with their nuclear cargo on planned manoeuvres to nearby Salisbury Plain. This is an action map, a battlefield map, like that of the Somme of 1916 (see p. 56), tailored specifically for one purpose.

American nuclear weapons had been placed in Britain to be nearer their possible Soviet Union target, part of an escalation of tension between the two superpowers during the 1980s. Britain's role, as the USA's equivalent of Russia's Cuba, was greatly facilitated by the 'special relationship' between Ronald Reagan and Margaret Thatcher. Nuclear weaponry, as with nuclear power, was a contentious issue for successive British governments. The Labour Party had learnt the hard way in 1982 that a policy of nuclear disarmament was not necessarily a vote winner, but while opposition to nuclear arms was meagre within Parliament, it was strong and articulate outside it.

CND was established early in 1958, with prominent figures such as the philosopher Bertrand Russell becoming members. As the production of such a map shows, CND was organised and well funded. Its members were overwhelmingly young, educated and middle class, with a popular left-leaning ethos that had been forged in the 1970s. Methods of non-violent protest included demonstrations, petitions and marches, but in the 1980s a new and unorthodox form of protest known as 'direct action' was introduced. Consisting of disruptive (and often headline-grabbing) behaviour, including vocal protest, obstruction and harassment, direct action was used by a number of environmental and feminist groups. Running parallel with the women-only peace camp outside Greenham Common, CND's direct action was named 'Cruise Watch'.

An accompanying printed note issued with the Cruise Watch map gives a valuable insight into its intended use 'to assist in locating, or following, and communicating if Cruise should drive south or south-west from Greenham'. In addition to the main roads along which the convoys travelled, the map provides secondary roads (tracks) and forest cover, which the instructions recommend be made more visible with a luminous pen. Most noticeable are telephone boxes – both public payphones and AA/RAC boxes. In an age well before mobile phones, call boxes presented the only means of effective and quick communication, apart from expensive walkie-talkies which were owned by only a few protestors. This was an area infiltrated by protestors, with their own sectors and knowledge of the landscape.

The convoys travelled at night, accompanied by bailiffs and police to their Salisbury Plain launch sites. When they were ambushed, they were attacked with paint and obstructed with intent to cause maximum aggravation and publicity. The obvious contradiction of direct action increasing the risk of danger to nuclear safety, the basis for CND's argument against them, did not prevent it from becoming a widespread tactic across Europe, especially in Germany, over the following decades.

1985: The guitar and the continent: the Live Aid logo

IVE AID WAS an international music event held on 13 July 1985 to raise funds and awareness for the Ethiopian famine of 1983–5. Some of popular music's biggest performers, including Madonna, Queen, David Bowie, Led Zeppelin, U2 and Bob Dylan, performed live at J. F. K. Stadium in Philadelphia, USA, and Wembley Stadium in London. The concerts were broadcast live in over a hundred and fifty different countries. This was a global event that harnessed satellite technology and mass media as forces for good. It had been BBC television reports and pictures of starving people – including the reporter Michael Buerk's description of 'the closest thing to hell on earth' – beamed to British homes, that first provoked Live Aid's co-founder, musician Bob Geldof, into action.

The concerts generated powerful images: the vast crowds, the superstars and the famous Live Aid logo of a guitar welded to a map of the African continent. A simple, smart visual pun, which captured the essence of the event, the logo was emblazoned upon two massive banners covering the amplifiers on each side of the Philadelphia and Wembley stages. It refocused the minds of the millions of viewers on the original purpose of the event. It was used heavily in merchandise, appearing on t-shirts and posters.

The logo played a crucially important role in the identity and communication of Live Aid through its cartographic simplicity. Maps have a long history as symbols. World maps known as T-in-O diagrams conveyed to medievalists the three continents of the world separated by a cross. Towns were represented on Roman coins by their most famous landmarks. Such necessary simplification, however, required sacrifices to be made elsewhere. This was the sacrifice: the Ethiopian famine had been caused by a mixture of drought, civil war and corruption in northern and southern Ethiopia, and neighbouring Eritrea. But many parts of Africa – Cairo, for example, or even the Ethiopian capital Abu Dhabi – were not blighted by famine. Identifying Ethiopia solely with famine – the iconic 'poor' country – was also deeply problematic, while not excluding apartheid-era South Africa in the logo constituted a missed opportunity.[131]

Altering the continental shape in keeping with the facts would have compromised the map's essential recognisable quality, but it associated all of Africa with famine, creating the generalisation that Africa *was* famine, a place defined by starvation. The association between the two was not only the doing of the Live Aid logo. The geographer Arno Peters' 1973 world map (upper right) used an

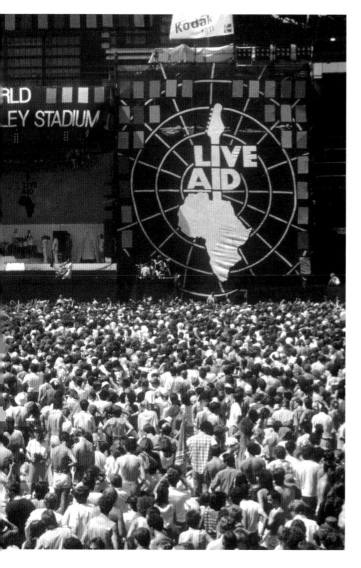

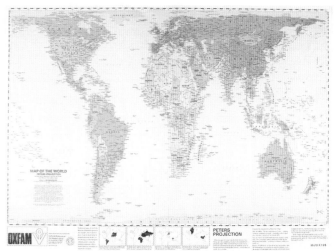

equal area map projection that made the equatorial zone, and Africa, appear far larger than in Gerard Mercator's projection of 1567, which Peters claimed distorted 'the picture of the world to the advantage of the colonial masters of the time'.[132] It was duly adopted by international aid agencies during the early 1980s. But it also distorted Africa's shape. And because instant recognition was the key function of the Live Aid logo, the design opted for the more familiar Africa of Mercator's projection. This was perhaps the ultimate riposte to Peters.

The musical instrument incorporated into the logo was an electric guitar, confirming Live Aid as a Western event for a predominantly Western audience. In fact, no traditional African music was performed on either stage on 13 July. By the 1980s the music industry was a global, multi-million-dollar concern. The left-wing radical protest music of Bob Dylan and others produced in the conscience-filled 1960s had been mass-produced at minuscule labour costs in free production zones. Cynics pointed to the fact that popular music with a conscience helped to sell more records; others highlighted the hypocrisy of the artists. Phil Collins, for example, flew on Concorde in order to play at both Live Aid venues, demonstrating just what was possible in the name of charity.

Although the Live Aid logo incorporated no direct reference to famine or starvation, images were repeatedly shown on live screens throughout the event, and a starving child was inserted into the updated version of the logo for the 2005 follow-up, Live Eight.

1986: Hunting with dogs: a map of landed Britain

THIS PASTEL-SHADED MAP showing the hunting districts of Great Britain could belong to an earlier age. The reclining gentleman with bugle and flat cap bears more than a little resemblance to the pipe-smoking gentleman studying the Ordnance Survey map on the popular edition cover of 1922. This nostalgia, of course, is part of the map's appeal. When it was produced in 1986, rural issues were well represented by a Conservative government. Hunting was a way of life and a popular sport throughout rural England and Wales. It had a strong tradition, an identity and geography of its own. Associated with the landed gentry, it nevertheless appealed to a wide range of participants and spectators, whose interests were reflected in the sorts of pictures and ornaments they chose to display in their homes. Hampton Editions' map was an artwork produced for this very enthusiast.

Here the hunting of rabbits and hares is shown using dogs, specifically beagles and harrier hounds, which tracked hares by scent. On the map each of the beagle and harrier packs' areas are shown roughly laid over county boundaries. They are geographically far more prevalent in southern England and the Midlands. The interest in hunting is strongly linked to the breeding of the dogs themselves: the Association of Masters of Harriers and Beagles, formed in 1891, endorsed this

map. The names of packs reflect the owners' military, public school and university education establishment backgrounds: Sandhurst, Eton College, Trinity Foot and Per Ardua, this last pack's name based on the motto of the Royal Air Force.

Hunts were spectacles, social and ceremonial occasions, which even served to assist the rural economy through the targeting and killing of wild animals regarded as a danger to farmers' livestock and crops. However, they were also highly controversial and subject to a number of legislative attempts at banning, most notably in the later 1990s. As a reaction, the Countryside Alliance formed in 1997 from the British Field Sports Society, but popular opposition to hunting eventually won through with a ban on hunting with dogs in 2004.

To many this was seen as a further erosion of the importance of the countryside. The geographer Danny Dorling's population map of 1995 (see p. 212), which shrunk proportionally less populated places down in size, symbolised the reduced importance of the countryside, especially to Parliament where far more votes were to be gained in urban areas. This was not ostensibly any different to the picture in 1986, though by the 1990s a Labour government had won the election with a pledge to ban what for its supporters was not the pastel-shaded world of the hunt, but the cruel and barbarous act of hunting.

1986: The growth of the supermarket: a map for expansion

SUPERMARKETS PLAYED a large role in altering the pattern and geography of everyday life in Britain. The first 'self-service' supermarket on the American model opened in Streatham, south London, in 1951, marking a real change to the traditional over-the-counter method of shopping in specialist shops.

This would compromise not only habit, but travel. Vast new supermarkets, or hypermarkets (defined as 25,000 square feet of sales area or over) started to crop up on the edges of towns where there was room, occupying land vacated by the disappearance of heavy industry in the late 1970s and 1980s. Customers had the space to park their cars, which became 'shopping baskets' to be filled up with the greatly expanded choice on supermarket shelves.

This laminated map from 1986 shows the locations of supermarkets across Britain. Clustered mainly around conurbations, these are coloured either red (for independent companies) or blue (for 'cooperative' companies). It is a strategic insider's map, produced by the Institute of Grocery Distribution (IGD) at a time when the number of supermarkets was rising steeply. Over the course of the 1970s, 162 stores were built; the same number again had been built by the mid-1980s. Forty new stores opened in 1986 alone, the year in which J. Sainsbury registered a pre-tax profit of £208.5 million and Tesco £122 million.

To assist in business expansion, the map provides a simplified topography free of natural or man-made obstacles, space (including designated green-belt land) to expand into. It shows the localism of major supermarkets: Tesco has a stronger presence in the south, its nearest rival Asda a stronger presence in the north, and Deefoods strong in Scotland. This localism would become less and less apparent as some supermarket chains grew, identifying new areas to move into, with computerised logistical systems improving supply chains and stock control.

In the volatile retail world, where name changes, rebranding and takeovers were common, companies may have had less reason to be forthcoming in contributing to market research such as this. The IGD map was accompanied by a comprehensive directory providing the complete lowdown on stores, extending even to the number of parking spaces and checkouts – apart from two. Sainsbury's didn't supply either detail and Tesco divulged only parking space numbers. Such reticence was perhaps understandable in a dog-eat-dog world.

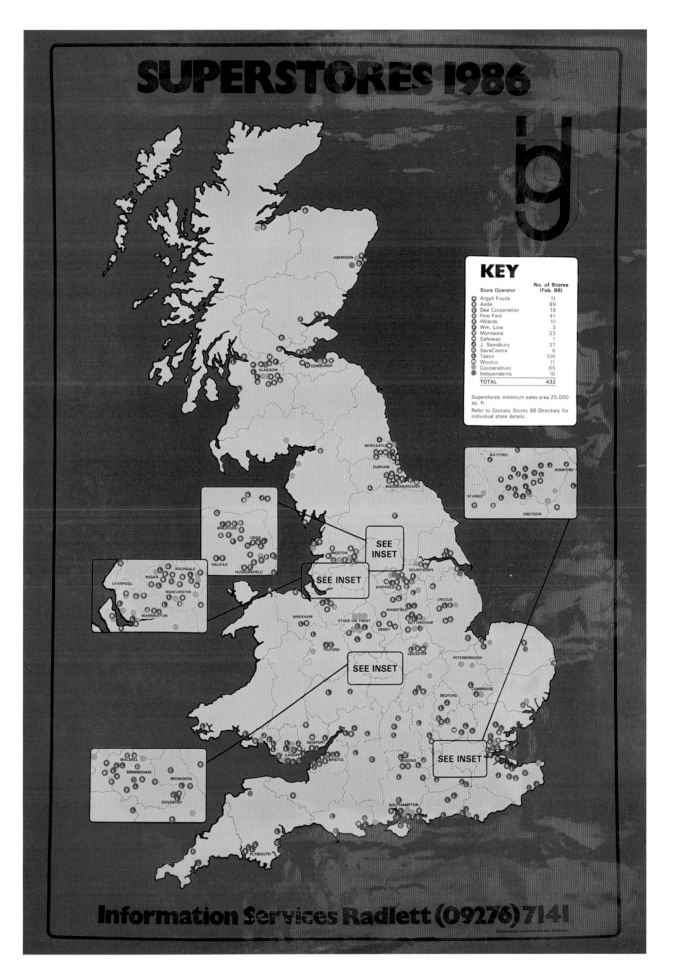

SUPERSTORES 1986

KEY

Store Operator	No. of Stores (Feb. 86)
A	Argyll Foods	11
B	Asda	89
C	Dee Corporation	19
D	Fine Fare	41
E	Hillards	10
F	Wm. Low	3
G	Morrisons	23
H	Safeway	1
J	J. Sainsbury	37
K	SavaCentre	6
L	Tesco	106
M	Woolco	11
	Cooperatives	65
	Independents	10
TOTAL	**432**

Superstores: minimum sales area 25,000 sq. ft.

Refer to Grocery Stores 86 Directory for individual store details.

Information Services Radlett (09276) 7141

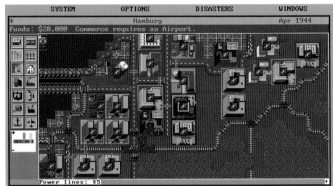

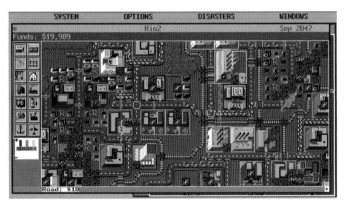

1989: SimCity: video games and urban planning for fun

B Y THE 1990s video and computer games were a lucrative global industry. Hardware and software companies in North America, Europe and Japan developed increasingly sophisticated and varied gaming experiences for a mainly juvenile audience. This, coupled with the increasing affordability of home computers and consoles, meant that millions of children started to stare at computer monitors instead of playing football or kiss-chase in the open air. The anti-video game lobby added moral danger to the health and obesity problems caused by video games, since in many of them the central goal was to kill as many enemies as possible.

It is easy to see how the mental stimulation and agility derived from playing these games, as well as the altruistic, educational features of some, may not have been appreciated by parents who had grown up in the 1960s. Yet two central features of video games, learning and developing skills and visualisation through maps – integral aspects of juvenile entertainment – came together in a video game of 1989 called SimCity.

SimCity was an urban-planning game written by the Californian programmer Will Wright and published by Maxis in 1989. Initially playable on Commodore and Macintosh consoles, the aim for the player was to create, build and run a city. Beginning with a computer screen plan of some uninhabited land and a bank account, the player created an urban infrastructure in stages: a power station, a road network, city zones, electricity and homes, which were subsequently populated by invisible 'Sims'. The player assumed the role of a mayor, imposing taxes, managing crises, keeping the Sims happy. Success was measured by popularity rating.

SimCity was unusual in that it was an open-ended game, without a narrative or discernible ending. Given how significant the crossover is here between 'work' and 'play', it is no surprise that Will Wright took inspiration from professional urban-planning scenarios and system dynamics. But what is really interesting for us about the game was that the player/mayor visualised his or her city by means of a screen dashboard of multiple maps. This was not merely a series of maps at various scales, but thematic maps – a crime map, pollution map, population density map – coloured to reflect particular aspects useful for the mayor's understanding. The map was the entire window on to the virtual world.

Accounting for the popularity of SimCity in 1989 (which has run to various later versions) we appreciate not only the compulsive and addictive nature of an apparently unending, infinitely complex game, but the pride of ownership, and the investment of the player's time in the creation of his or her own city. We can also identify a confluence between the virtual and real worlds in the game, for SimCity, as with all maps, betrays its origins.[133] Beginning a city from scratch, and creating it as an ordered grid pattern, is a style of city building which is particular to the United States.

It could readily be argued, of course, that the player did not start with 'nothing', but a natural environment, with trees, grass, earth and some coastline. The assumption for player and designer was that the building of a city was a positive thing, the only sensible thing to do. The capitalist, corporate, urban vision of the Thatcher–Reagan years was alive and well, in contrast to the counter-cultural movement of the 1960s, which had argued rather strongly against it. But better to be done virtually, in a computer game, than in real life? It would be for the following century's virtual communities to agonise over which was truly real.

1990: Tribal maps in Belfast

ETAILED STREET PLANS of Belfast (and other parts of Northern Ireland), shaded to reflect religious distinctions and generically known as 'tribal maps', were widely used by the security forces, including the Army and Royal Ulster Constabulary, from the 1970s onwards. For speed and ease of association, the predominantly Catholic residential areas were coloured green, Protestant areas were orange and the make-up of mixed areas could be roughly calculated by the hue, so that yellow reflected a fifty–fifty mix. Commercial or industrial areas were left white. This type of map was not originally military (similar maps illustrated the suppressed Irish Boundary Commission Report as far back as 1925), but the Army quickly came to appreciate the value of this rudimentary means of assessing the likely levels of threat in a neighbourhood.

These examples, covering the Lower Falls and Ballymurphy Turf Lodge areas of west Belfast, were cut down and laminated in the early 1990s, probably for use in joint Army/RUC vehicle checkpoint duties. They were often clipped to the commander's board inside military vehicles such as the Snatch Land Rover (developed for use in Northern Ireland in 1992), but may also have been worn on a map pad on the thigh, to be read in a crouching position. 'Tribal' maps were issued in various other forms, from large, general maps that were displayed in operations rooms, to the highly detailed area maps that were cut to the shape of the stock of the SLR rifles then in use. Once taped in place they could be read at a glance by simply tilting the weapon while carrying the rifle in the cradle position (in shoulder but with the muzzle lowered, considered less threatening for urban policing duties than carrying weapons at the ready). These versions were often laminated with a soft, cling-film-like substance that turned slowly brown until the map became unreadable but didn't cut the cheek if the rifle was deployed.

The purpose of these cut-down maps was to make map-reading as swift and inconspicuous as possible. Small foot patrols were sent out on to the streets of Belfast throughout the three decades of the Troubles and, as a former Army officer noted, 'The old, golden rule of any combat is get the unit commander – the guy with either the map or the radio (or both). So discreet use of either was prudent'.[134] Newly deployed officers and NCOs who paused at junctions to unfold conventional maps were easy targets, and their evident uncertainty about their whereabouts also made them vulnerable. The security forces worked on the assumption that IRA snipers knew their business and would line up firing marks or lines of sight at street corners, a foot or so below the street signs, at a height where a bullet would strike the head or chest; when patrol leaders began to crouch down in order to read maps the IRA cells rapidly adapted. The safest practice was to keep moving and to look confident about location. Large-scale maps showing detail of individual houses and gardens – maybe limited to a dozen streets in all – were also essential in sealing off an area in the event of an incident.

These ephemeral maps have not been manipulated for any form of political point-scoring. They could only fulfil their function if scrupulously accurate. More than any other type of document, 'tribal' maps throw into stark relief the self-imposed religious segregation at all levels of society and at all stages of life which underpins Northern Irish politics. For almost half a century after the Irish Civil War (1922–3) politicians of all stripes in Belfast, Dublin and London deferred taking any direct action that would upset the status quo. The thirty-year cycle of atrocity and counter-atrocity, which began in the late 1960s, was finally halted by the Good Friday Agreement in 1998 as a consequence of general weariness and revulsion, rather than through any fundamental demographic or ideological shift. While the Republic of Ireland was riding high on the Celtic Tiger, at least, extremists of both camps came to be regarded as 'bizarre antiquarian relics of another age', irrelevant in an era of 'post-national Euro-prosperity'.[135] The truce has held, but the riots and demonstrations that greeted the decision to restrict the flying of the Union flag over Belfast City Hall in December 2012 are a reminder of its fragility.

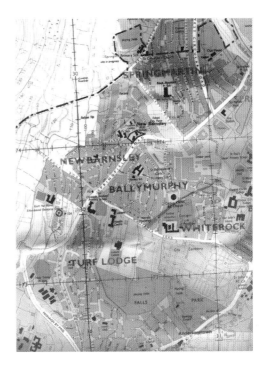

1991: Public and secret maps of the Gulf War

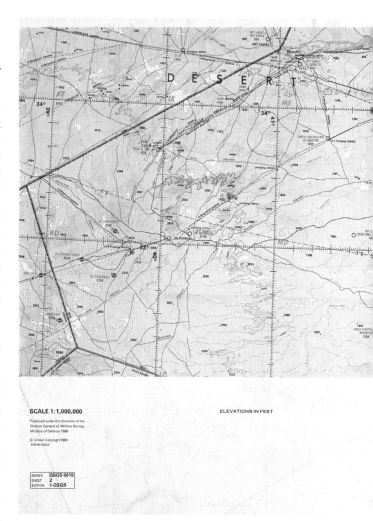

SCALE 1:1,000,000   ELEVATIONS IN FEET

Produced under the direction of the
Director General of Military Survey,
Ministry of Defence 1990

© Crown Copyright 1990
8/90 8619/2/2C3

SERIES	GSGS 5619
SHEET	2
EDITION	1-GSGS

THE GULF WAR (1990–1) was notable for its unprecedented media coverage, earning it the title of the first televised war. Live news pictures and audio of the US-led coalition attack upon Iraq, in response to Saddam Hussein's invasion of Iraq's oil-rich neighbour Kuwait in August 1990, were broadcast around the world from the front line and from the bombed Iraqi capital Baghdad.

Following the Iraqi invasion, and a period of diplomacy and sanctions, coalition forces, including those of Britain and Saudi Arabia, moved in. Air strikes attacked military and government targets, including some in Baghdad. These bombing raids began on the night of 16 January 1991 and were reported live by Western media correspondents based in the city. When the bombing raids disrupted live picture feeds, the American CNN was the only network able to continue live broadcasting in audio.

In the absence of pictures, a digital map graphic of the Middle East appeared on the screen, with an inset photograph of the correspondent. Baghdad, the location of the report, was the only place marked. The map provided a very basic function as a cartographic screensaver while the live action continued. As the standard of these graphics improved towards the end of the century, so they became ever more interactive with the news reports.

Vastly more precise maps were required by armed forces engaged in the bombing operations. Accuracy was especially important in order to avoid civilian casualties in highly populated areas such as Baghdad. The Royal Air Force joined the coalition in the aerial assault of Operation Desert Storm. This 'Escape and Evasion' map of 1990 was compiled from operational navigational charts to assist air operations, and is printed on Paxar, a deliberately durable material which is also used for the laundry labels of clothes.

Escape and Evasion maps remained part of aircrew survival kit, as they had done since their introduction in the Second World War. Opportunities for using them in the relatively flat and cover-free terrain of Iraq were limited, and aircrew increasingly relied on their personal locator beacons to attract Combat Search And Rescue teams (CSAR).

Six RAF Tornado aircraft were lost in combat during the first Gulf War, their crews killed or captured. The first to be shot down was crewed by John Nichol and John Peters. Captured and paraded on Iraqi television, their bruised and beaten faces became enduring images of the war, another intersection between this conflict and the media. Nichol, the navigator, described the fate of the 'route map' he was carrying. Fearing that the CSAR reference points marked on it might be of value to Iraqi intelligence, he 'ripped out a bloody great chunk of it, stuffed it in my mouth, and started chomping ... the rest of the map I buried'.[136]

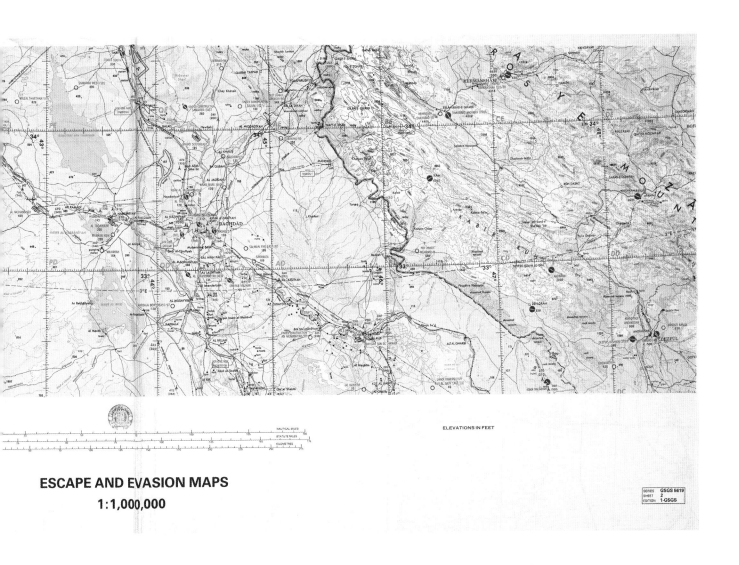

ESCAPE AND EVASION MAPS

1:1,000,000

ELEVATIONS IN FEET

SERIES GSGS 5619
SHEET 2
EDITION 1-GSGS

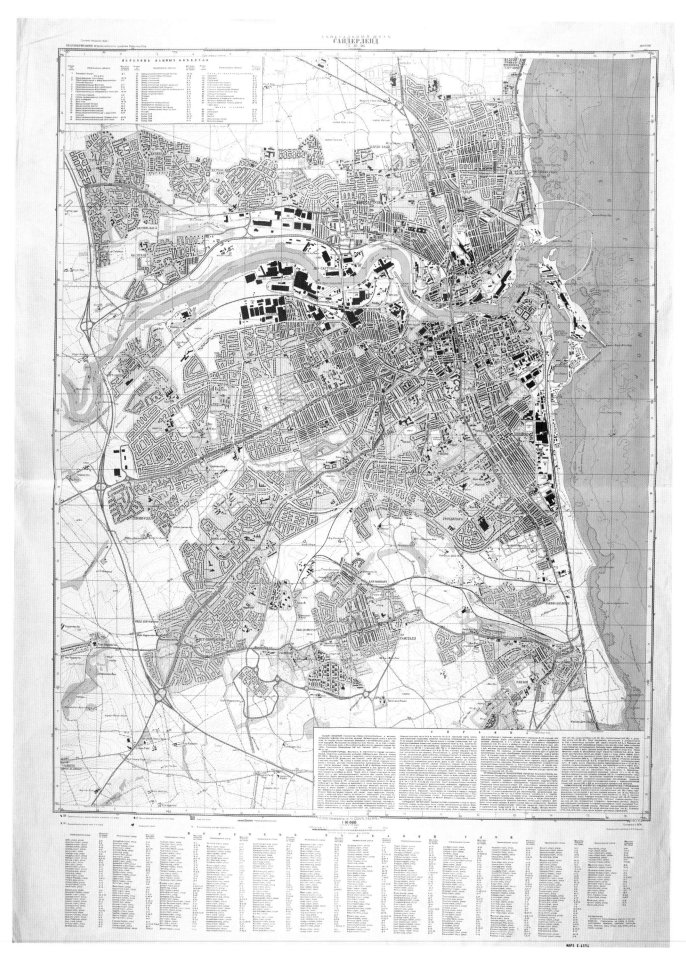

1992: Soviet Sunderland: part of a global mapping project

THE COLOSSAL GLOBAL mapping programme of the Soviet Union and her allies was uncovered after the fall of communism in 1992. Following the opening up of Eastern Europe, official Soviet military mapping began appearing for sale in Europe and America, a large quantity (measured in tonnes) apparently from abandoned military map depots in Latvia and other former Soviet states. It comprised topographical paper maps at a range of seven scales, with a world-wide classification system, and included thousands of urban street plans, of which at least a hundred were towns in the British Isles. Some of them had been updated several times over the years.

As we have seen throughout this book, one does not necessarily need to have jurisdiction over somewhere in order to map it. Once executed, such mapping can facilitate actual – and symbolise psychological – control over its subject. It is fair to say that these maps terrified authorities and experts when they were revealed, not because a Soviet invasion was imminent in 1992, or might have been in the past (currently the earliest known urban map of this type is dated 1938), or even as a result of lingering fears about communism. It was because the maps were apparently as good as maps produced in the UK.

The Soviet map of Sunderland in north-east England, made in 1976, exhibits these qualities and production values. It is a large, densely packed lithographic sheet, entirely in Cyrillic script, with a gazetteer and legend. The map has been printed in ten colours, with individual buildings coloured to reflect their use. By comparison, Ordnance Survey mapping at the same 1:10,000 scale contains only two colours and monochrome block buildings. Ordnance Survey would argue in 1993 that the Russian mapping had breached their copyright, but there was far more to it than simple plagiarism. Too much of the information was different.

Such unusual familiarity with these and other parts of Sunderland, including even some allotments in Southwick, may have been gleaned by spies ten years earlier. The Soviet Union's participation in the 1966 World Cup finals provided fantastic cover for reconnaissance, and the Soviet national team played two first-round games and their quarter-final tie at Sunderland's football ground, Roker Park. A recent study of Russian maps has concluded that they were compiled from:

> everything from high technology to low skulduggery. Satellite images and high-altitude aerial reconnaissance played a major part, but there was also a diplomat (who was not a cartographer) stationed in every Soviet embassy round the world whose job it was to collect all possible information, by fair means or foul. All available published maps, guides, directories and similar documentation were gathered and dispatched to Moscow. Where these did not suffice, then illegal means such as bribery, theft or blackmail were used. Money, it seems, was no object.[137]

The precise purpose of the massive Soviet project has not been satisfyingly established. Surely such a colossal undertaking had a clear point? It has been suggested, though surely tongue-in-cheek, that 'every Soviet president from Stalin to Gorbachev ... not only knew where you lived, but how to get there by tank'.[138] For generations that had grown up in the 1960s with spy planes, spy satellites and John le Carré novels, there was a thrill and excitement in the voyeurism associated with Brezhnev knowing what you had eaten for breakfast. The answer is probably the same as the reason behind the continuous production of Ordnance Survey mapping: for information, just in case the need for it arose. The hard truth for post-colonial Britain was that, even when on the receiving end, Britain was no longer special, merely one part of a global mapping project.

1994: Walt Disney World:
the suburb where dreams come true

WALT DISNEY WAS a highly successful animator and entertainer, capable of turning make-believe into reality, as this tourist map of a modern-day utopia proves. Disney Corporation's films embody (not necessarily universally held) values of freedom, hard work and entrepreneurial enterprise approximating to the 'American dream'.

Disney was also a shrewd businessman, conceiving a place where families could actually step into the magical world of his films. Walt Disney World in Florida opened its doors in 1971. The second theme park to have been constructed along these lines, it is a large residential holiday resort containing attractions, merchandise and experiences, a triumph of enjoyment and mass consumerism. Set around the centrepiece fairytale castle, it includes a funfair, water parks, the 'Magic Kingdom', and the EPCOT (Experimental Prototype Community of Tomorrow) Center, Disney's urban vision of the future. It also harks back to America's past in 'Main Street USA', a recreation of an early 1900s Midwest high street.

By 1994, Walt Disney World had welcomed its 400 millionth visitor. It was cleverly located in Orlando, which became not only the American holiday destination of choice, but that of millions of international visitors from Europe and Asia, even after the opening of Disney resorts in Paris, Hong Kong and Tokyo. In order to cater for this audience, the American map publisher Rand McNally produced a special tourist map of the Greater Orlando area, for sale in the UK and Europe. It shows, among the suburbs of Kissimee and St Cloud, the suburb of Walt Disney World, which appears not merely as a theme park, but as a real place.

And so it was. Disney World was established as a city, with city boundaries, power of jurisdiction and even its own town mayor. When compared to places such as Las Vegas or even Prince Charles' model English town of Poundbury (see p. 153), Walt Disney World is equally authentic. Indeed, as (mostly) French theorists have pondered while writing about Walt Disney World's utopian representation of America, it is all too real.

Rand McNally had been publishing tourist maps for years. Their 1929 pictorial chart of English literature had been designed for the American tourist travelling to Britain (see p. 74). Although in their 'City Flash map of Orlando' the roles are reversed, we see in the earlier portrayal of sites such as Stratford-upon-Avon and historic London the same mixture of myth, reality and pastiche as is present in their map of this Orlando suburb.

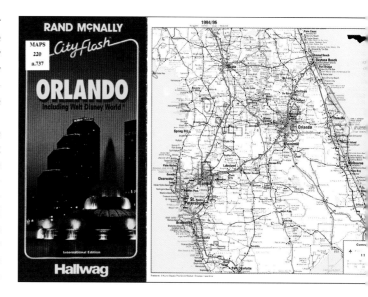

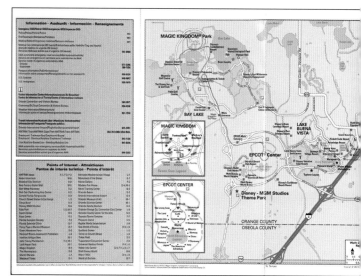

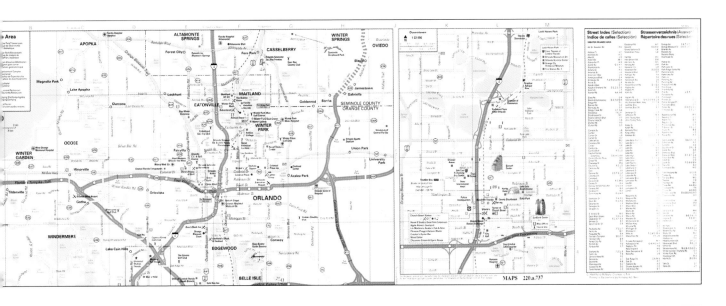

MAPS 220.a.737

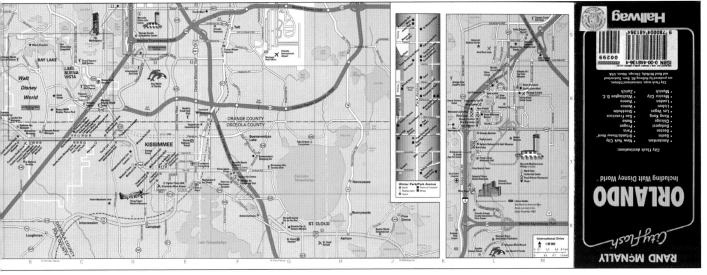

OUT OF THE RED AND INTO THE BLUE 1973–1999 **207**

1995: A map of Internet traffic: information beautiful and useful

WE ARE TODAY accustomed to seeing the Earth as a sphere suspended in space. Even before the astronauts of Apollo 8 photographed the globe for the first time in 1968 (see p. 154), successive civilisations had been anticipating what it looked like with varying degrees of accuracy.[139] After 'Earthrise', the image of the globe became even more prevalent and was reproduced on flat surfaces and in three dimensions, animated in moving images and recreated as simulations on computer screens.

This image is a frame taken from such a computer animation. It was used to illustrate a paper given at a conference on 'Information Visualisation', though in the animation the globe was viewable from a variety of angles, and the arcs connecting points on the globe with each other were also dynamic.[140] These arcs are not the paths of missiles, though as has been pointed out, they share similarities.[141] What they show is the flow of Internet traffic, online communication between capital cities from 1 to 7 February 1993. The application was created by scientists at the telecommunications company Bell-Labs in Illinois, USA. Its purpose was to make usage statistics between Internet servers visible in order to understand them better.

We've already seen how visualising statistics cartographically can provide a more informative, though not necessarily more accurate, picture (see the 1945 map showing the occupation of women by regions, p. 110). This animation was able to add and subtract statistics, using colour, intensity, height and transparency to articulate every aspect of Internet activity for the period under study. Like any modern road system, the flow of online 'traffic', which suffered the same bottlenecks and accidents as actual road traffic, required constant maintenance. Such visualisations served a practical purpose in assisting scientists' analysis and improvement of the service.

The Internet had its origins in United States government studies into transferring secure data during the 1960s, but the programme funded by the National Science Foundation (NSF) in the 1980s, and the NSFNET backbone, was developed to link the main academic computing centres in North America. The date of Bell-Labs' animation is significant because by 1995 the Internet had become a commercial service for public use.

Rapidly increased traffic (16 million users in 1995 leapt to 248 million in 1999)[142] necessitated sophisticated maintenance equipment. But the value of this image moves well beyond its original purpose of the programme for the Bell-Labs engineer/motorway repair person. It provides the general geographic location of Internet users (grouped into major cities), giving some global picture of access to computer hardware. It also shows the connections that users were making. Lines (or arcs) of communication link migrants in North America and Europe with their Latin American homeland.

This animation is an artefact from the early years of the Internet and World Wide Web, which by the end of the century had developed such a rich and complex topography that it had obtained its own atlas.[143] Its short history had become interesting. In addition to practical, historical and academic angles (the academic conference on the then young field of data visualisation), wonder at the complex beauty of information visualisation became an art form in its own right.

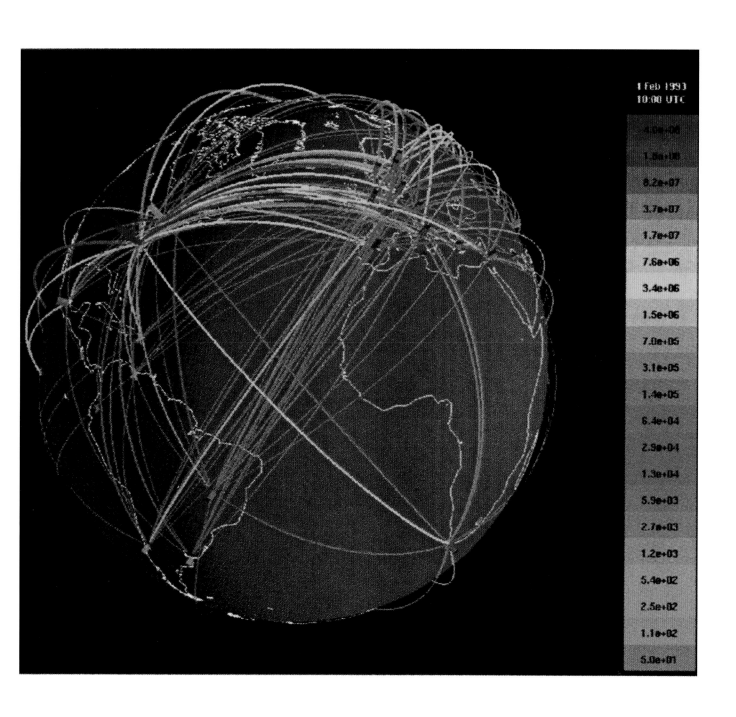

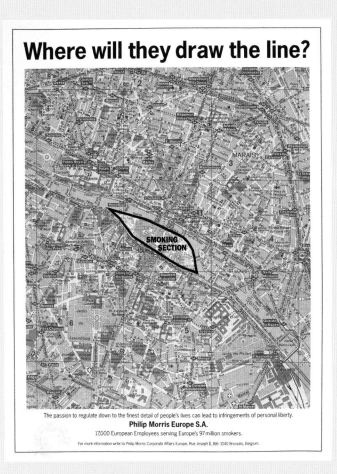

Where will they draw the line?

The passion to regulate down to the finest detail of people's lives can lead to infringements of personal liberty.

Philip Morris Europe S.A.

17,000 European Employees serving Europe's 97 million smokers.

For more information write to Philip Morris Corporate Affairs Europe, Rue Joseph II, 166-1040 Brussels, Belgium.

Where will they draw the line?

The passion to regulate down to the finest detail of people's lives can lead to infringements of personal liberty.

Philip Morris Europe S.A.

17,000 European Employees serving Europe's 97 million smokers.

For more information write to Philip Morris Corporate Affairs Europe, Rue Joseph II, 166-1040 Brussels, Belgium.

1995: Where will they draw the line?
Tobacco advertising and the death of the smoker

CONSIDERING THEIR POWERFUL subconscious communicative abilities, it is surprising that maps were not used more in tobacco advertising. From the now-faded giant adverts on the sides of walls, still whispering slogans such as 'for your throat's sake smoke', to such subliminal creations as the purple Silk Cut calendar, tobacco advertising throughout the twentieth century attracted the largest budgets and the most inventive marketing minds. Tobacco advertising was a £25 billion UK industry in 2002, but in that year it became illegal in the European Union. The final Silk Cut advert contained a purple-clad fat lady singing.

A sense of the incredulity about the imminent EU smoking ban is visible in these adverts, placed by the South African tobacco giant Philip Morris in the *Economist* magazine in June and July 1995. They show the city centres of Paris and London, each with a black border drawn around small areas of streets labelled 'smoking section'. The virtual 'sealing off' of these areas is a clear reference to the increasingly powerful anti-smoking lobby in the EU, and legislation intended to ban smoking outright in pubs and restaurants. It was eventually introduced in Britain in 2007 and France in 2008. The irony of this advert is that even cafés and bars in the Ile de la Cité or Soho would not for long be enclaves for smokers.

Britain had always smoked. But the tide began to turn in the 1980s with a ban on smoking on public transport, and by the late 1990s the weight of medical evidence and opinion was firmly against it. Anti-smoking groups upped the ante, inverting tobacco advertising's themes. American icon Marlboro man's cigarette drooped, turning masculine appeal on its head by indicating a link between smoking and impotence.

This led to a backs-against-the-wall mentality in the pro-smoking lobby. In Germany in 2000, where a voluntary ban on smoking did not have the desired effect, a t-shirt produced by a pro-smoking group, emblazoned with a Star of David and the word 'Raucher' (smoker), was swiftly banned. Philip Morris's advert was a call-to-arms, an angry riposte to the public health warnings, powerfully evoking an 'us versus them' mentality of a group who were, or who at least saw themselves as, stigmatised.[144] This was not the initiative of a pro-smoking group, but a tobacco manufacturer who profited from its users' self-stigmatising activity. Philip Morris urged smokers of the world to unite.

The maps' accompanying statement ratcheted up the pressure: 'The passion to regulate down to the finest detail of people's lives can lead to infringements of personal liberty.' By drawing boxes around small, central and (from the impression of the map) cramped areas, the implication was that smokers themselves were confined to these tiny places, deprived of their freedom and forced to live in concentration camp-like conditions. Yet if these havens were the only place they could smoke, smokers would doubtless willingly have taken it.

Such symbolism could, however, have adverse effects. The line on the map heightened the sense of alienation and undesirability felt by smokers, even in the eyes of the occasional smoker, or the non-smoker with smoking relatives. It separated smokers from their allies, and polarised opinion. In the end, the British ban on smoking in public places passed off with few newsworthy moments, as smokers were pushed outdoors and ostracised.

1995: *A New Social Atlas of Britain*: turning statistics into pictures

NCONVENIENTLY FOR THE human geographer, Britain's population is not evenly spread. People do not occupy the same amount of space and, as a result, it isn't easy to illustrate the incredible amount of data and statistics collected through surveys such as the decennial national census. The issue is one of scale. In Britain, as with a great many other places in the world, a few tiny urban areas contain the majority of the population. To the human geographer's mind, Manchester is actually bigger than Northumbria because more people live there.

By 1995, the technology, software and programs existed to create a map in which a scale of miles could be replaced by a scale of people. The geographer Danny Dorling presented his new map in the groundbreaking *New Social Atlas of Britain*. On the left is 'recognisable' Britain and on the right is the equal area cartogram of Britain. Where the scale of the former is a 1,000-kilometre-square box, the cartogram's scale is 250,000 people.

The highly populated areas – London, Birmingham, south Wales, the central belt of Scotland – are distinctly swollen, while mid-Wales, north Devon and the Scottish Lowlands appear as limp hanging skin. It is an unpleasant analogy, but the similarity to the human body is appropriate. Dorling's bloated bagpipe-shaped map is the reality of human Britain.

The *New Social Atlas* visualises 106 statistical data sets on population, demography, economics, housing, health, society and politics – issues interesting to human geographers, social historians and government departments. This map shows the proportion of ethnic-minority residents living in Britain in 1991. It is an important statistic, reflecting on a very basic level (any person who ticked a box other than 'White' on the census form) the degree to which Britain had become more multicultural after 1945. In the cartogram, areas with between 0.1 and 5 per cent ethnic minority residents are shaded in greys. Areas with between 5 and 25 per cent are pink.

The map contains nothing more than the supplied data, and as such can only be as good as the data it expresses. But a picture can be worth a thousand words. Digitally enabled data visualisation and human cartography of this kind became extremely valuable in allowing people to interpret figures, and see patterns and trends that are not as clearly discernible in tables or numbers. Here the cartogram not only enables comparisons of minorities in cities, but their arrangement within those cities.

It also helps that information visualisation could be beautiful and appreciated increasingly as an art form for its complex scientific elegance. Such beauty could influence the viewer's perception of the information, and the patterns contained therein.

Creating a map of Britain in which everyone was equal served as a way of redressing the inequality existing in British society. In fact, in its use of bold pink for the minorities, it could be argued that this map marginalises the majority. By the mid-1990s, a perception that the policies of over fifteen years of Conservative government had widened the gap between rich and poor provided the circumstances for such statements to be made.

Ethnic Minority Residents 1991

proportion of ward populations

% of residents in ethnic minorities

up to 0.1%   13

0.1% to 1%   73

1% to 5%   11

5% to 25%   2

25% & above   0

% of the land area of Britain

Scale

☐ = 1000 km²

% of residents in ethnic minorities

up to 0.1%   1

0.1% to 1%   39

1% to 5%   37

5% to 25%   17

25% & above   6

% of all residents in Britain

Scale

☐ = 250 000 people

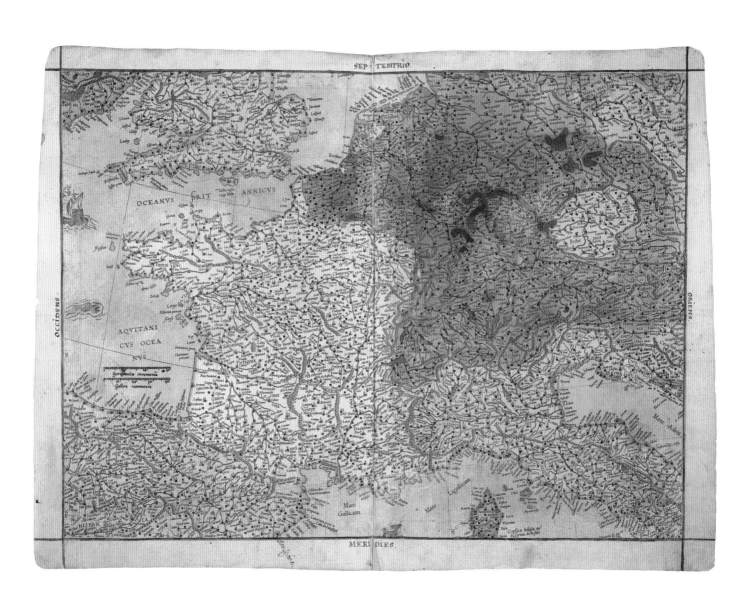

1996: The *Mercator Atlas of Europe*: old maps as investments

THE *MERCATOR ATLAS* of Europe was compiled in around 1570 by the mathematician Gerard Mercator as a gift for his patron, the Cleves court official Otto von Gymnich. It becomes relevant to our history on 26 November 1996, when it was auctioned for sale at Sotheby's in London. Its then owner was British Rail, which had purchased it in 1979 with £340,000 of their workers' pension fund. This is an old map in the guise of an investment in the era of privatisation.

British Rail, from 1948 the publicly owned national railway company of Great Britain, was privatised in 1997. Throughout the 1980s, in common with other nationalised industries, it was placed under pressure by reduced government financial support. It had thousands of unionised employees paying into extremely good state pension schemes. Rather than simply safeguarding these funds by hiving them off into sideline businesses and private companies, sound investment could also bring significant return. Investment in art and antiques had become a credible alternative to stocks and shares, especially in the United States, where New York supplanted London as the centre of the art world. Association with culture was itself a commodity and prices for modern art hit astronomical heights.

British Rail enlisted the help of Sotheby's auctioneers in amassing a portfolio. From the mid-1970s, with an economic backdrop of 17 per cent inflation and the world oil crisis, it put around 2.5 per cent of the pension fund into art and collectables, by 1980 having invested £8.8 million on 2,132 items.[145] It was a diverse portfolio, including not only art but Chinese porcelain, furniture, collectables and cartography. This highly unusual strategy attracted some criticism and accusations of bad investment. The purchase of a rather tatty 400-year-old atlas, not exactly on the same scale as a Renoir or a piece of Chippendale furniture, was a notable example of what must to some have seemed a bizarre acquisition.

Historical maps and atlases have been collectable antiques for many decades, the main attraction often being their artistic appearance. But beautiful this atlas was not. It was a rough-and-ready collection of hand-drawn and printed, cut-up, annotated maps, bound in paper covers with a scribbled title. This unassuming appearance would explain why it remained unknown until discovered in a Belgian bookshop in 1967. Maps are also valued for their rarity or significance: those believed to have changed the course of history or to have been the 'first' to show something have been known to attract colossal amounts at auction. Even tatty maps have value, as the Dean and Chapter of Hereford Cathedral realised in 1988 when their medieval mappa mundi was proposed to be auctioned by Sotheby's, the reserve price £3.5 million.[146]

The financial value of this particular atlas lay in its unique creation by the hand of Mercator, the man responsible for naming a book of maps an 'atlas' for the first time and devising a world map on which sailors could plot a straight course. If any historical map-maker's name was recognisable to the majority it was Mercator's. As an investment, it is easy to appreciate why British Rail 'persistently refuse[d] scholars' requests for close inspection',[147] keeping it under strict access conditions in the National Library of Scotland – despite the fact that mention was made in a 1980 board report of 'special provision to be made for access to study specific items'.

As scholars and, crucially, enough senior administrators realised, the atlas had research as well as monetary value. Amid questions over the quality of British Rail's investments and ethical issues over what it was proper for the company to possess, the atlas was put up for sale. After failing to reach the reserve price, it was eventually sold in May 1997 for well in excess of £500,000. The buyer was the British Library, a government-funded organisation, with the aid of a large grant from the National Lottery Heritage Fund.

1997: The geography of mourning:
a newspaper maps the public funeral of Diana, Princess of Wales

NUMEROUS CARTOGRAPHIC montages appeared in British newspapers on the morning of 6 September 1997. Main maps of central London showed the funeral procession route of Diana, Princess of Wales, to be made later that day. Other maps showed the area around Althorp House, Northamptonshire, the Spencer family seat, where Diana's coffin was to be driven afterwards for burial. A few even included an inset of south-east England, giving the relation of Althorp to London and the M1 motorway linking them.

This geography of mourning was combined with photographic illustrations of the funeral venue Westminster Abbey and likely attendees, with headlines such as 'Route of memories', 'Diana 1961–1997: the final journey' and 'The people's princess: a funeral to reflect Diana's place in our hearts'. The maps fulfilled a different function to the analytical 'scene of the crime' graphics of the car crash in the central Paris road tunnel that had killed Diana, her companion Dodi Fayed and their driver six days earlier.

These pages functioned as news souvenirs and practical guides for those not only wishing to line the streets along the route or watch the funeral on television, but also those who perhaps preferred to avoid it. The exhaustive media coverage of 'mass mourning' in the days following Diana's death ignored the fact that as many people did not actively mourn as placed bouquets of flowers in their thousands at the gates of Kensington Palace. The episode stimulated a number of scholarly articles in cultural journals, and formed the plot to a successful film which portrayed the then only recently elected Labour prime minister Tony Blair as the true perceiver of public opinion.

Both the palace and, initially, the *Sun* newspaper were seen to have misjudged the strength of public feeling. Diana, of the aristocratic Spencer family, had married Prince Charles, heir to the English throne, in 1981. She was young, glamorous, as at ease with famous personalities as with the average person on the street. But her marriage, the birth of her two sons and 1996 divorce were all played out under the scrutinising glare of the media. The black car in which she died crashed while being chased by paparazzi press photographers on motorbikes. Her brother, the 9th Earl Spencer, said in a live press conference, 'I knew the press would kill her in the end'.

The involvement of the press in the life and death of Diana continued in her commemoration. Her state funeral (and a half-day of mourning, though a Saturday) was granted by the palace under the weight of public pressure. The event was anticipated and comprehensively covered in the press. The newspaper maps may have influenced some to gather beyond London, lining the A1 trunk road and M1 motorway. People threw flowers at the hearse; on a number of occasions the driver had to stop for flowers to be cleared from the windscreen. Whether or not the press were responsible for Diana's death, the event tapped into a powerful psyche which gave expression to feelings of loss, sadness and grief – all of it extremely newsworthy.

FINAL JOURNEY

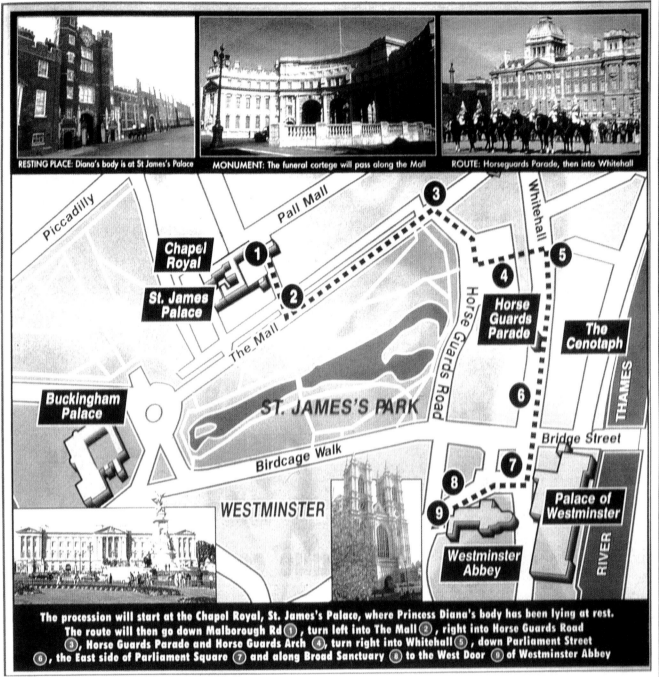

RESTING PLACE: Diana's body is at St James's Palace   MONUMENT: The funeral cortege will pass along the Mall   ROUTE: Horseguards Parade, then into Whitehall

The procession will start at the Chapel Royal, St. James's Palace, where Princess Diana's body has been lying at rest. The route will then go down Malborough Rd ①, turn left into The Mall ②, right into Horse Guards Road ③, Horse Guards Parade and Horse Guards Arch ④, turn right into Whitehall ⑤, down Parliament Street ⑥, the East side of Parliament Square ⑦ and along Broad Sanctuary ⑧ to the West Door ⑨ of Westminster Abbey

day of the funeral. Flags will be flown at half mast.

State funerals are usually for sovereigns. But on rare occasions they have been arranged to honour exceptionally distinguished people such as Sir Winston Churchill.

There is also a ceremonial royal funeral for members of the royal family with high military rank.

A third category is a private royal funeral for other members of the family, their spouses and their children.

The decision to give Diana a Westminster Abbey ceremony follows a Mirror campaign to win back her HRH status.

The Mirror wrote to the Queen's Proctor, who represents the crown in matrimonial matters, urging him to intervene in the royal divorce

proceedings and make Diana Her Royal Highness again.

We asked our readers to sign a petition to express their outrage.

Now the palace has bowed to public feeling. And as Diana lies dead, her place in the nation's life will be officially recognised.

● DIANA'S brother Earl Spencer arrived in England from his home in South Africa yesterday and signalled his family's full support for the planned people's funeral.

The 33-year-old earl said it was "right and proper" for Britons to have the chance to pay their respects.

He added: The family is grateful that Buckingham Palace is making every effort to accommodate its wishes at such a sensitive time.

ROUTE OF MEMORIES

By ROGER WILLIAMS

THE funeral route will bring back many poignant memories of Diana's life.

The CHAPEL ROYAL at St James's Palace, where her coffin now lies before the altar, is where she and Charles joyfully witnessed the christening of Andrew and Fergie's first daughter Beatrice in 1988.

Four years later both couples were to separate.

Diana's coffin will be borne along THE MALL, where she and Charles rode in their carriage through huge cheering crowds to Buckingham Palace after their wedding in July 1981.

Memories of a month earlier that year will be evoked as the funeral procession goes down HORSE GUARDS ROAD. It was the day the then Lady Diana Spencer went there to watch Trooping the Colour, a man fired blanks at the Queen and the bride-to-be made her first appearance on the balcony of Buckingham Palace.

PARLIAMENT STREET and WHITEHALL house the offices of many of the Establishment figures the People's Princess railed against as "them." And PARLIAMENT SQUARE is where then Premier John Major arrived to announce her marriage split.

On to WESTMINSTER ABBEY – scene of triumphant royal coronations and sombre funerals through the centuries.

Diana's most prominent visit was in 1986, for the ill-fated marriage of the Duke and Duchess of York.

TEARFUL HEWITT'S TRIBUTE: PAGE 11

1997: The Glastonbury Festival: the image, the reality and the Criminal Justice Act

ADMISSION TO THE three-day Glastonbury Festival of Contemporary Performing Arts of June 1997 was £75 plus booking fee. 'Thank you for supporting us by buying your ticket' gushed organiser and founder Michael Eavis in the official programme's introduction. Ninety thousand people paid, perhaps as many again didn't. Yet even those who jumped the perimeter fence would have been able to buy the event programme published by *Select* magazine, which described the greatest line-up of popular music acts anywhere on earth that weekend.

The programme's centre spread contained a cartoon map of the festival site entitled 'Where are you now?'. This light-hearted caricature map is interesting in that it captures the tension between the idyllic and free ideology behind the festival's image, and the requirements of an unmanageably vast, increasingly corporate event.

The modern free festival movement originated with the development of a counter-culture in the 1960s, of which music was a crucial element. The Woodstock Festival in 1969 and Windsor Great Park events organised by Ubi Dwyer from 1970 were followed that same year by the first Glastonbury Festival. Never technically free, though embodying aspects of these earlier events, Glastonbury was the brainchild of a Somerset farmer (Eavis) and located in the mystical Vale of Avalon. It grew over the following thirty years to become the country's largest entertainment festival.

The image of an alternative, non-commercial, nature-embracing event was understandably difficult to maintain by the 1990s, though Glastonbury retained an ecological charitable edge compared with similar but more unashamedly commercial festivals. A percentage of the ticket sales went to organisations such as Greenpeace and Oxfam. Hints of this alternative, leftist ideology are embedded in the cartoon map in the official licensed beer tents. These were run by the trade union-owned 'Workers Beer Company', and bars were given names such as

'The Mandela' and 'Rainbow Warrior', after the Greenpeace ship which was sunk by French intelligence services in 1985 with loss of life.

The map contains a few fallacies. Aside from the profusion of grassy fields (1997 was to become known as 'the year of the mud'), it represents the festival site as a quaint garden, when in fact it was a sprawling temporary city of at least ninety thousand people. The rickety wooden fence stands in great contrast to the succession of large metal perimeter fences, the space in between patrolled by security guards with dogs. The characters populating the map are gently mocking festival 'types', including the traditional New Age traveller and their 1990s equivalent, the 'raver'.

These were the lifestyles that perpetuated the counter-culture tradition so central to the image of the festival. They were also lifestyles affected by the 1994 Criminal Justice and Public Health Act. A Conservative government law, the legislation criminalised the 'gathering on land in the open air of 20 or more persons (whether or not trespassing) at which amplified music is played during the night', giving special note to music 'characterised by the emission of a succession of repetitive beats'. The legislation was designed to protect rural landowners but the implications for people's freedom, and their civil liberties, were wide-ranging.

Tension between the authorities and followers of the rave scene, very often travellers, had been present as early as 1985 when police attacked travellers attempting to access the Stonehenge site for a festival. The police were later found guilty of wrongful arrest, criminal damage and assault, but the new law gave them greater power. Dance act The Prodigy (whose singer is included on the Glastonbury map as they performed on the Friday evening) were particularly vocal in their opposition to 'their law'. The map of the leafy garden showed not only a world away from the Glastonbury Festival behemoth of the 1990s, but something beyond what was possible in Britain without government permission.

1999: A cartographic chronicle of the fall of Yugoslavia

THE WARS COMPRISING the break-up of Yugoslavia were a regular feature in Western news bulletins and newspapers throughout the 1990s. They used various maps and graphics in their ongoing coverage of the overlapping conflicts in Slovenia, Croatia, Bosnia and Kosovo, particularly the four-year siege of Sarajevo by Bosnian Serb forces (1992–6) and the Serb massacre of refugees at the United Nations safe haven of Srebrenica (1995). Yet, despite the proliferation of media mapping, as well as copious, conflicting maps establishing sovereign territory and enforcing treaty boundaries, no official, irrefutable and cumulative 'base' map of what was happening was kept, even at the United Nations.

The task was taken up by FAMA International, a Sarajevo-based multimedia enterprise founded by Suada Kapic and her son Miran Norderland in 1990 during the instability caused by the fallout of crumbling communist institutions. According to its website, FAMA had intended to become the 'CNN of south-eastern Europe', and its work intensified in the wake of wars begun by Slobodan Milošević, president of Serbia, the dominant constituent state of Yugoslavia, towards achieving an ethnically homogeneous Greater Serbia.[148] This process had escalated through declarations of independence by Slovenia and Croatia (in 1991) and Bosnia (in 1992). Serbs living in these states carved out autonomous enclaves, assisted from Belgrade with force, and met with opposition by national armies and paramilitary groups.

Besieged in Sarajevo, FAMA collected images, testimonies and oral histories, produced survival guides and documentaries, and held festivals and exhibitions. Its 'Sarajevo Survival Map 1992–1996' was drawn in a basement in the town and smuggled out through an underground tunnel. After the war this material was placed in educational study packs and encyclopaedias, and in 2012 was digitised as part of the Virtual FAMA collection. 'The Fall of Yugoslavia 1991–1999', a map charting the war's entire nine-year span, was proudly dated June 1999, the month which marked the end of NATO air strikes that had pummelled Serbia into final submission over the Albanian refugee massacre in Kosovo. It was a large, colourful, glossy poster in English, with a vast chronology (beginning with the 1980 death of Yugoslav president Tito) and mini city dossiers describing the fate of different cities in Croatia and Bosnia Herzegovina.

The achievement of the map lay in its 'success in presenting an issue that usually takes 1,000 pages – on just one sheet of paper'.[149] Miran has explained how it was motivated by a request from a United States congressman (one of many international visitors in Sarajevo following the siege) to provide a background brief on the Yugoslav conflict in five minutes. Nine years is long for a war by twentieth-century standards, but condensing its history and presenting it cartographically helped to explain it succinctly to the wider world, as well as facilitating some form of acceptance for its survivors. This is cathartic cartography, but it pushed the capabilities of the map to its limit.

Produced using Corel and Photoshop software programs, and printed in Slovenia, the map included text boxes and a fifty-seven-piece legend explaining the symbols of national flags, bombers, fleeing refugees (the symbol for ethnic cleansing), paramilitaries, and burning buildings and tanks. Out of all the elements, Miran is particularly proud of the chronology on the reverse. It was, he explains, the fruit of painstaking collaborative research, and he is incredulous that such a task had not been attempted elsewhere. Need for legibility was the reason for the map's large size. But such was the stretching of the image required to achieve the necessary dimensions that the appearance of the map, particularly the originally hand-drawn symbols and decorative corners, has the distorted quality of a video game graphic.

Through its bold straightforwardness the map achieved its educational aim, as well as generating the funds for FAMA to continue their work documenting the period and promoting their mission of reconciliation in the region. It had a wide audience, in addition to map shops and map libraries, proving a popular souvenir for the SFOR (UN mandate stabilisation force) soldier from Idaho, and an indispensable point of reference for the news reporter and the diplomat to take home. An example of the map was given to the then US president Bill Clinton, and a copy hangs in the offices of the International Criminal Tribunal for the former Yugoslavia in The Hague.

1999: Millennium tapestry map: Britain's communities

A CROSS THE UK, a large number of tapestry and embroidered maps were produced by local communities to commemorate the millennium. They provide a counterpoint to sophisticated digital mapping and aerial photography (see *The Photographic Atlas*, p. 227). The tapestries were not commercial ventures; they were made possible thanks to small grants including ones provided by the Millennium Commission. By contrast to the digital reproduction and printing processes of much contemporary cartographic output, they were produced using traditional, labour-intensive methods. And whereas projects like the *Photographic Atlas* were very consciously national in their scope and coverage, the millennium tapestries celebrated the 'locality'.

They were unique hand-crafted objects, placed on display at the centres of the communities that made them (usually the village or church hall) and tell a story of people and places that doesn't quite fit with twentieth-century words such as urbanism, progress and advancement, but which still found expression through a map. The word map, incidentally, derives from the Latin word 'mappa', meaning 'cloth'. Needlework had fallen out of favour for the postwar generation, but there was a revival in the 1980s with prominent work including the NAMES Project AIDS Memorial Quilt, and Afghan-made tapestries initiated by the Italian artist Alighiero Boetti.

Various millennium projects had been established in Sherington, Buckinghamshire, to celebrate the village's past, present and future. An embroidery map of the village, initiated by a lady called Enid Pepper and formed of a series of patchwork squares, would represent Sherington's present, but the finished embroidery's colourful pattern of fields, buildings and other features shows a broader historical spread. For example, much of the building of the church,

St Laud's, is over 800 years old. Other buildings have had a range of uses, such as the old chapel which was by then a home, and there is a red telephone box, by then something of a rarity even in rural England. Farming and school activities take place. Had the embroidery been larger, there might have been visible, through the window of a certain house, ladies sitting around a table stitching the replicas of their own houses. Although Mrs Pepper does not believe the embroidery is a 'proper' map because it is not made to scale, it is in many ways as authentic a representation of Sherington as was possible to make.

Rural life had changed over the century. Few would realise from the embroidery that a large new town had been constructed a few short miles away from Sherington (see p. 152). Change could balance itself out. For example, many rural places in the UK had been isolated by the railway cull of the 1960s (see p. 120). Sherington had, however, benefited from the nearby M1 motorway. During the final decade of the twentieth century the closure of rural amenities such as post offices and banks, and the disappearance of local shops, were offset by large out-of-town shopping centres and the internet marketplace.

Yet many elderly people would not or could not adapt to these new life patterns. Not everybody owned a car. They relied on their neighbours, on state-funded services, on their communities for help. The number of retired and elderly people in Britain had never been higher by the year 2000, partly thanks to improvements in healthcare. The Sherington embroidery is a reflection of its makers, many of whom had lived through the majority of the twentieth century and who had witnessed at first hand many of the events described in this book. The map snapshot incorporates past and future, change and continuity.

Postscript

I: The end of the century: how Britain viewed itself and its neighbours

The cartoon strips of *Viz* magazine were rude versions of the *Dandy* or *Beano* comics, which had been staples for British schoolboys from the late 1930s. Founded in Newcastle in 1979, *Viz* peddled a sexual, violent and highly offensive brand of humour through the travails of characters such as 'Sid the Sexist' and 'The Fat Slags'. Unsurprisingly for these very reasons it was highly popular, with sales exceeding a million copies during the 1980s. Ironically, from being published on a card table in the founder's bedroom, *Viz* was absorbed into the establishment in 1987 when the new owner, John Brown, appointed his father (Sir John Brown, former 'Publisher' or chairman of Oxford University Press and board member of the British Library) as a director.

In 2001, *Viz* produced a special pull-out supplement map of Europe, on which its artist had been free to create a no-holds-barred picture of European contemporary culture as seen from Britain. It contained those Continental stereotypes that had been percolating in the British psyche for many years, exhibited in some form in the early board game *Trip to the Continent* (see p. 16) as well as the serio-comic 'Hark! Hark! The Dogs do Bark' (see p. 46). It presented the attitudes of the average British person, probably male, towards Britain and Europe.

Amid the violence, sexual activity and demonstrations of bodily functions depicted on the map are references to specific events. In the British Isles we see sectarian tension in Northern Ireland, the famous 'north–south divide' debate, and the long-running feud between Spanish and British fishermen. In France, pill-popping cyclists refer to doping scandals that dogged the famous Tour de France, while dishevelled Disney characters paint a seedy picture of the Disneyland Paris resort. French cuisine does not escape, nor does relaxed legislation on recreational drug use in Holland. Scandinavia is filled with people committing suicide, while Eastern Europe is a luminous green colour, referring to the Chernobyl disaster of 1986 (see p. 168). The power station is joined by the rusting Russian Mir space station, which crashed back to Earth in 2001.

This hand-drawn and hand-coloured poster would have adorned many a bedroom wall, reinforcing the occupants' stereotypes, clarifying what they had seen on holiday and in the news, and what their parents had told them. Its cultural value has already been noted: in an exhibition on British comic art at the Tate Gallery in London in 2011, *Viz* artwork sat together with the famous satirical cartoons of William Hogarth and George Cruikshank.

Placed together, their take on the world was not so different. Such is the pervading vision of Europe in portions of the British media that the largely insular British audience's one-dimensional view of it was never likely to be nuanced. The experience of Europe was acquired off the television or through a holiday. Yugoslavia? A big nasty war. Spain? Bullfighting and holiday apartment blocks. Germany? Fat sun-bathers, Adolf Hitler and Maria von Trapp from the 1960s musical *The Sound of Music*. Many of these were easy targets, but the map exhibits some perceptive humour too. McDonalds restaurant logos appear over some of the most historical European monuments. For the illustrators of *Viz*, perhaps even some aspects of contemporary life had overstepped the line.

II: Air photography: the perspective of half a century

We have already seen how, as the end of century and millennium drew nearer, the need to commemorate them found expression through maps (see p. 222). *England: The Photographic Atlas* was published by the Millennium Mapping Company and HarperCollins using the aerial imagery of the company Getmapping plc. It was the sort of landmark, unwieldy cartographic object that was concerned as much with symbolism as with usefulness (the vast *Klencke* and *Earth Platinum* atlases, of 1660 and 2010 respectively, were even more colossally impractical). But it had a special significance in that it contained the first comprehensive photographic record of England.

Over seventy thousand images, taken from an aeroplane at 5,500 feet, were digitally manipulated into photographic mosaics. The result wasn't just a millennium 'snapshot' of the country: it demonstrated the range of technical and publishing expertise achievable by the year 2000. It also showed the degree of access to aerial imagery that the general public was by then permitted, which in previous years would have been declared 'sensitive' on security grounds.

In 2014 satellite and aerial photographic imagery is extensively available. Google Earth offers coverage of the entire world online, and the most rural places of the UK, Canada and Siberia are reasonably well covered. But in 2000 it was significantly more difficult to find detailed aerial imagery anywhere other than in books, and then it was not always possible to specify any location one liked.[151]

Fifty years earlier, aerial photographs by the Royal Air Force were published by the Ordnance Survey as expedient and temporary alternatives to actual maps (photographs constituted a stage in the mapping process, topographical detail being transferred from photo to map by photogrammetry). These large black-and-white images showed the level of devastation wrought by enemy bombing during the Blitz and assisted post-war reconstruction by government departments. From 1948, to recoup costs, the mosaics were offered up for public sale, but in 1950, in response to unspecified security intelligence, they were doctored to obscure sensitive sites such as airfields. In 1954 their sale was stopped completely.[152] Such glorious coverage of south-eastern and urban Britain would certainly have been handy to any state contemplating a vast undercover mapping project as the Cold War drew in (see Soviet Sunderland, p. 204).

Threats to national security had not substantially diminished by 2000 (they were to substantially increase the year after), yet public

access to aerial and satellite imagery such as *England: The Photographic Atlas* had increased. Perhaps something is still hidden: though the luscious imagery may be perfectly open and digestible, the infrastructure supporting it remains hidden and ultimately under the control of government agencies.[153]

If anything had been purposefully excluded from the photographic atlas, in addition to the whole of Wales, Scotland and Northern Ireland, it may have been illegible anyway. Such were the size requirements for the atlas that the rural areas were shown at too small a scale to recognise properly. Roads, far narrower on a photograph than in a map, are no aid to navigation. *The Photographic Atlas* was a large, complicated and difficult object to produce, a doorstop to the twentieth century rather than an opening to the twenty-first.

III: Around Ground Zero

The benchmark provided by the turn of the century has been supplanted by the benchmark of 9/11. Recent history is more commonly understood as occurring either before or after the events on 11 September 2001 in New York City, where hijacked passenger aeroplanes crashed into and destroyed the World Trade Center, one of the world's tallest buildings, killing over 2,800 people. These events were witnessed by millions on live television. The Islamist militant organisation Al-Qaeda claimed responsibility for the attacks. In response, the United States launched a 'war on terror' against Afghanistan in 2001 and Iraq in 2003.

To many the world changed as a result of 9/11. But in the immediate aftermath of the destruction of the Twin Towers, there was tragedy, confusion and a vast hole in the ground at the heart of one of the world's biggest cities. Shortly afterwards a map was produced of this devastated area of Lower Manhattan that had been christened Ground Zero. It provides an important insight into how a map may not only facilitate a practical and emotional navigation of a place, but explain how the two are intertwined. Its creator, Laura Kurgan, has written of it:

> The map was produced as both a practical guide to the site and as a memorial document. It addressed the multitude of people who were going to the site to see it for themselves, people who were simply looking or who were seeking to add something of their own to the many spontaneous memorials that had proliferated there. The aim of the map was to help people make sense of what they were seeing or, if that was asking too much, at least to measure their disorientation in the face of the imaginable. The site around the World Trade Center was manifestly disorientating, but the map sought to address that confusion and allow visitors to begin to take stock of what had happened.[154]

We have seen this same dual practical-emotional mapping in the ceremonial processes following the death of Diana (see p. 216) and also in the work of the artist Richard Long (see p. 158). The use of maps to assist virtual pilgrimages to sacred sites of any description is one of their oldest functions.

One of the main purposes of this book has been to uncover the powerful ability of maps to document events as well as the places where they took place. Often, though not always, maps outlive both. As the scene of destruction in New York became a building site, and replacement towers were constructed, this map became a crystallised version of the site's former reality, available for excavation by later generations curious to know what had happened on that fateful day, or looking to explain the reasons behind what took place afterwards.

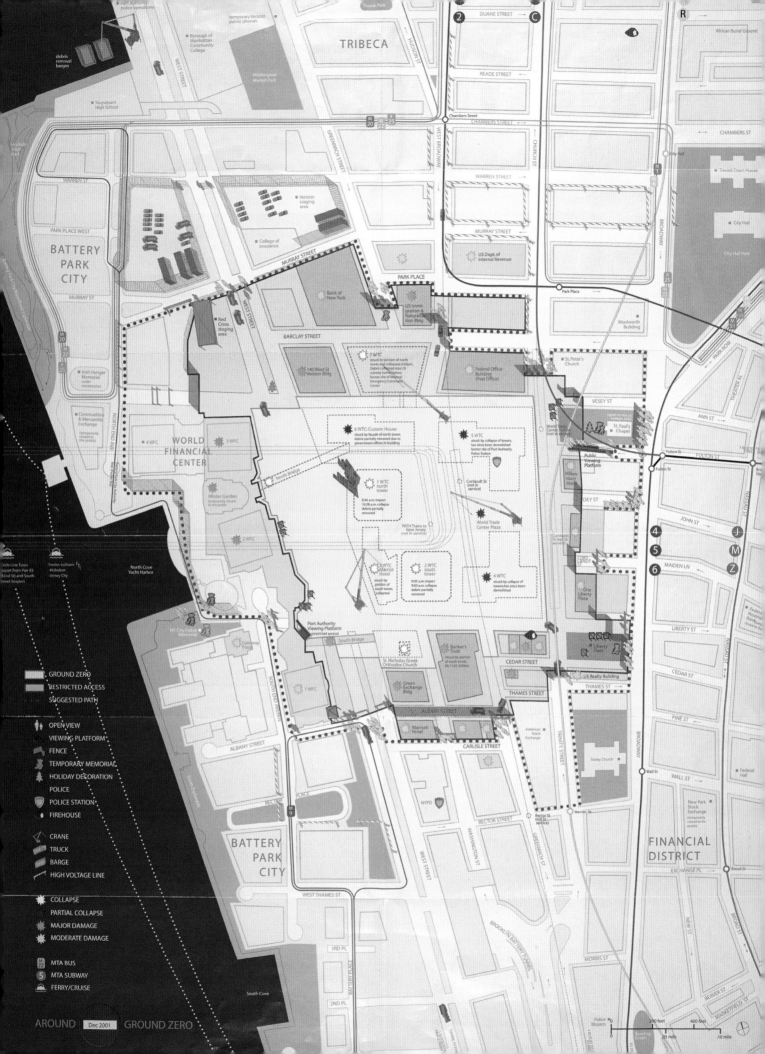

Endnotes

1   A London underground poster identified by Felix Driver shows how one could visit sites of empire simply in the capital city. Felix Driver and David Gilbert (eds), *Imperial Cities: Landscape, Display and Identity* (Manchester: MUP, 1999), p. 1.

2   Catherine Delano-Smith and Roger J. P. Kain, *English Maps: A History* (London: British Library, 1999), p. 1.

3   Graham Greene, *Our Man in Havana* (London: Heinemann, 1958), p. 7.

4   See for example J. B. Harley and David Woodward's preface in Harley and Woodward (eds), *The History of Cartography volume one: Cartography in Prehistoric, Ancient and Medieval Europe and the Mediterranean* (Chicago and London: University of Chicago Press, 1987), p. xvi.

5   Ibid, p. 24.

6   See P. D. A. Harvey, *The History of Topographical Maps: Symbols, Pictures and Surveys* (London: Thames & Hudson, 1980), p. 86.

7   We are grateful for insights provided by Christopher Board.

8   Walter Kursten, quoted in http://www.europe-ana1914-1918.eu/en/contributions/6487.

9   James R. Ackerman, 'Finding Our Way' in James R. Ackerman and Robert W. Karrow Jr. (eds.), *Maps: Finding our Place in the World* (Chicago: University of Chicago Press, 2007), p. 19.

10 Roland Barthes, *Image, Music, Text* (London: Flamingo, 1977), p. 18.

11 Karen Cook, 'The Historical Role of Photomechanical Techniques in Map Production' in *Cartography and Geographic Information Science*, 29:3 (2002), pp. 137–154.

12 Tim Owen and Elaine Pilbeam, *Ordnance Survey: Map Makers to Britain since 1791* (Southampton: Ordnance Survey, 1992), p. 118.

13 Michael Thompson, *Rubbish Theory: The Creation and Destruction of Value* (Oxford: Oxford University Press, 1979).

14 For a sample of the debate see René Larsen and Dorte V. P. Sommer, 'Facts and Myths about the Vinland Map and its Context' in *Zeitschrift für Kunsttechnologie und Konservierung* 23:2 (2009), pp. 196–205.

15 Gary Magee and Andrew Thompson, *Empire and Globalisation: Networks of People, Goods and Capital in the British World, c. 1850–1914* (Cambridge: Cambridge University Press, 2010), p. 241.

16 *Pygmalion*, Shaw's most popular play, was first performed in Vienna in 1913; productions opened in London and New York in 1914 and separate editions were published by Constable (London) and *Everybody's Magazine* (New York).

17 Jerome K. Jerome, *Three Men on the Bummel* (Bristol: Arrowsmith, 1900), p. 133.

18 See Richard Mullen and James Munson, *The Smell of the Continent* (London: Macmillan, 2009).

19 Jerome, op. cit., p. 314.

20 Roger Hellyer and Richard Oliver, *Military Maps: The One-inch Series of Great Britain and Ireland* (London: Charles Close Society, 2004), p. 5.

21 For these figures and the suggestion about the left-handed binder, see Christopher Board, *"Certainly better than nothing at all": A Re-examination of the Imperial Map of South Africa, 1899-1902* (unpublished paper presented at the 21st International Cartographic Conference at Durban in 2003).

22 We are indebted to historian Elizabeth Van Heyningen for the suggestion about location; see also her book, *The Concentration Camps of the Anglo-Boer War. A Social History* (Auckland Park: Jacana, 2013).

23 C. Russell and H. S. Lewis, *The Jew in London: A Study of Racial Character and Present day Conditions* (London: T. Fisher Unwin, 1900); it has also been discussed by Peter Barber in *London: A History in Maps* (London: British Library, 2012), pp. 236–7.

24 See V. D Lipman: *A History of the Jews in Britain since 1858* (Leicester: Leicester University Press, 1990).

25 For a full discussion see Rosemary O'Day and David Englander, *Charles Booth's Inquiry: Life and Labour of the People in London Reconsidered* (London and Rio Grande: Hambledon Press, 1993).

26 Prakash Shah, *Refugees, Race and the Political Concept of Asylum in Britain* (London: Cavendish Publishing, 2000), p. 37. The Act was put to the test in the aftermath of the pogroms following the 1905 Russian Revolution, which generated a fresh wave of asylum seekers. According to Shah, the only grounds for exclusion noted, across the board, were want of means and medical grounds. Asylum seems to have been largely granted.

27 Bernard Porter, *Plots and Paranoia: A History of Political Espionage in Britain, 1790–1988* (London: Unwin Hyman, 1989).

28 Kenneth O. Morgan, 'The Boer War and the Media (1899–1902)' in *Twentieth Century British History* (vol. 13:1, 2002), p. 7.

29 See Julia Gillen and Nigel Hall, *The Edwardian Postcard: A Revolutionary Moment in Rapid Multimodal Communications* (paper presented at the British Educational Research Association Annual Conference Manchester, 2–5 September 2009).

30 J. G. Bartholomew (ed.), *Atlas of the World's Commerce* (London: George Newnes [1907]).

31 Andrew Porter, *The Oxford History of the British Empire: Volume III: the Nineteenth Century* (Oxford: Oxford University Press, 1999), p. 404.

32 Jerry White, *London in the Nineteenth Century* (London: Jonathan Cape, 2007), p. 150.

33 For a full discussion, see *Marek Kohn, Dope Girls: The Birth of the British Drug Underground.* (London: Lawrence & Wishart, 1992).

34 'Relief map modelling' in *The Geographical Teacher*, vol. v, 28:6 (Autumn 1910), pp. 303–14.

35 An extra volume in the series, entitled 'The soldier's

geography of Europe', was specially prepared for the use of soldiers in training and sold at cost price (3 d.), published from 1910.

36 Huigh A. Dempsey, *The CPR West: The Iron Road and the Making of a Nation* (Vancouver and Toronto: Douglas & McIntyre, 1984).

37 Dempsey, op. cit., p. 200.

38 John A. Eagle, *The Canadian Pacific Railway and the Development of Western Canada, 1896–1914* (Montreal: McGill-Queen's University press, 1989), p. 249.

39 Proposing the toast to the British Empire at the annual Madras Law Dinner, April 1915. M. K. Gandhi, *Collected Works* (New Delhi: Publications Division) vol. 14, pp. 417–18.

40 Quoted in Alfred Lansing, *Endurance: Shackleton's Incredible Voyage* (London: Weidenfeld & Nicolson, 1959), p. 11.

41 Peter Chasseaud and Peter Doyle, *Mapping Gallipoli* (Staplehurst: Spellmount, 2005), p. 321.

42 *British Empire Union Monthly Record* article reprinted in New Zealand, Jan 7 1919 edition of *Akaroa Mail and Banks Peninsula Advertiser* (retrieved from http://paperspast.natlib.govt.nz/cgi-bin/paperspast ?a=d&d=AMBPA19190107.2.12.

43 André Chéradame, *The Pangerman Plot Unmasked* (London: John Murray, 1916).

44 Sir Stuart Campbell, *Secrets of Crewe House* (London: Hodder & Stoughton, 1920).

45 See Sean McMeekin, *The Berlin-Baghdad Express* (London: Allen Lane, 2010).

46 Peter Chasseaud, *Artillery's Astrologers: A History of British Survey and Mapping on the Western Front 1914-1918* (Lewes: Mapbooks, 1999).

47 Martin Middlebrook, *The First Day on the Somme: 1 July 1916* (London: Penguin, 1971), pp. 179–82.

48 J. B. Post, *An Atlas of Fantasy* (Baltimore: Mirage, 1973), p. 90.

49 Ricardo Padron, 'Imaginary Worlds' in Akermann and Karrow (eds), *Maps: Finding our Place in the World* (Chicago: University of Chicago Press, 2007), p. 278.

50 The *Morning Post* also serialised the *Protocols of the Elders of Zion*, an unpleasant dig at Montagu, reported to be more sympathetic to 'asiatics' because he was Jewish.

51 John Martin Robinson, *Queen Mary's Dolls' House: Official Guidebook.*(London: Royal Collection Enterprises, 2004). Also http://www.royalcollection. org.uk/queenmarysdollshouse/house.html.

52 Julian Symons, *The General Strike: A Social Portrait* (London: Cresset Library, 1987), p. 25.

53 Ann Thwaite, *A. A. Milne: His Life* (London: Faber and Faber, 1990), p. 298.

54 Thwaite, op. cit., p. 522.

55 John Bilingsley, *The Day the Sun Went Out: The Background to the 1999 Total Solar Eclipse: Accounts of the 1927 Eclipse as Seen from Yorkshire and the Pennines* (Hebden Bridge: Northern Earth, 1999), pp. 7–8.

56 John Paddy Browne, *Map Cover Art* (Southampton: Ordnance Survey, 1991), pp. 90–1. The British Library example, as with all popular maps, has had its covers cut off and disposed of. This standard British Museum practice illustrates how maps were valued for their data and nothing more.

57 Derby *Daily Telegraph*, 21 May 1927.

58 This would continue through to the end of the century, this very map being used as an exhibit in the 1999 Library of Congress show *John Bull and Uncle Sam: Four Centuries of American–British Relations*.

59 Sara Haslam, 'Wessex, literary pilgrims, and Thomas Hardy' in Nicola J. Watson (ed.), *Literary Tourism and Nineteenth Century Culture* (Basingstoke: Palgrave Macmillan, 2009), p. 165. For Americans' experience of Europe, see from the same source Shirley Foster, 'How America inherited literary tourism', especially p. 175.

60 Geoff King, *Mapping Reality* (Basingstoke: Macmillan Press, 1996), p. 3, after the *Guardian*, 1990.

61 Dov Gavish, *The Survey of Palestine under the British Mandate 1920–1948.* (London: Routledge Curzon, 2005), p. 239.

62 Gavish, op. cit., p. 238.

63 John Hope Simpson, *Report on Immigration, Land Settlement and Development in Palestine* (London: HMSO, 1930), p. 56.

64 Paul Moon, *New Zealand in the Twentieth Century: The Nation, One People* (Auckland: HarperCollins, 2011), p. 198.

65 A. Tocker, 'The effects of the trade cycle in New Zealand' in *Economic Journal* 24 (1924), p. 128, quoted in Moon, op. cit., p. 185.

66 Julie V. Gottlieb, *Feminine Fascism: Women in Britain's Fascist Movement*, (London: I.B. Tauris, 2003).

67 Diana to Unity 17 September 1936, quoted in Diana Mosley (ed.), *The Mitfords: Letters between Six Sisters* (New York: Harper Perennial, 2008), p. 76.

68 Geoffrey J. Giles, 'Student Drinking in the Third Reich', in Susanna Barrows (ed.), *Drinking: Behaviour and Belief in Modern History* (Berkeley: University of California Press, 1991), p. 135.

69 Hans Ulrich Thamer, 'The Orchestration of the National Community: The Nuremberg Party Rallies of the NSDAP', in Gunter Berghaus (ed.), *Fascism and Theatre: Comparative Studies on the Ethics of Performance in Europe 1925–1945* (Oxford: Berghahn Books, 1996).

70 Though at that time groundbreaking lithographic mapping had been produced by the Quartermaster General's Office.

71 Alaric Searle, *Ideology and Total War: Military Intellectuals and the Analysis of the Spanish Civil War in Britain* (http://usir.salford.ac.uk/11474/3/MGZ_SCW_short_version.pdf).

72 As a comparison, the earliest printed Ordnance Survey map of Kent, 1801, simply omitted any detail of the naval dockyard at Chatham, leaving a prominent blank space on the map.

73 Burl Burlingame, *Advance Force: Pearl Harbor* (Annapolis: Naval Institute Press, 2002), p. 281.

74 Explored fully in Colin Smith, *England's Last War Against France: Fighting Vichy 1940–42* (London: Weidenfeld & Nicolson, 2009).

75 The artist, Blake, trained at Camberwell School of Art but had been working as an architectural draughtsman. His stint as a war artist for the Ministry of Information opened new doors for him after the war as a successful commercial artist and respected painter.

76 See Anthony Beevor, *D-Day: The Battle for Normandy* (London: Viking, 2009).

77 Richard Nelson Gale, *With the 6th Airborne Division in Normandy.* (London: Sampson, Low, Marston & Co, 1948), p. 87.

78 Gertrude Williams, *Women and Work* (London: Nicholson & Watson, 1945).

79 The clearest modern account of the events behind this futile gesture has been assembled by Anthony Beevor in *Berlin: The Downfall, 1945* (London: Viking, 2002).

80 For example by Tony Judt in *Postwar: A History of Europe since 1945* (London: Heinemann, 2005).

81 Miles Taylor, *Southampton, Gateway to the British Empire* (London: I.B. Tauris, 2007), p. 181.

82 A. C. Merton-Jones, *British Independent Airlines since 1946, Volume I* (Liverpool: Merseyside Aviation Society, 1976).

83 Sydney Clarke, *All the Best in Europe* (New York: Dodd, Mead & Co., 1955), p. 58.

84 Roger Bray and Vladimir Raitz, *Flight to the Sun: The Story of the Holiday Revolution* (London and New York: Continuum, 2000).

85 The Queen herself had been presented with a special map display which showed, by means of lights and switches, the position of SS *Gothic* on any day (stated in typewritten handout no. 340/53, Admiralty, 5 November 1953, issued to personnel).

86 Figures on publication history taken from http://www.tolkienbooks.net/php/lotr-print-runs.php. Sources include the Allen & Unwin Archive at Reading University.

87 Ricardo Padron, 'Mapping Imaginary Places' in James R. Ackerman and Robert W. Karrow Jr (eds), *Maps: Finding our Place in the World* (Chicago: University of Chicago Press, 2007), p. 273.

88 Martin Drout, *J. R. R. Tolkien Encyclopedia: Scholarship and Critical Assessment* (London: Routledge, 2007), p. 408.

89 Richard Oliver, 'Episodes in the history of the 1:25,000 map family' in *Sheetlines: The Journal of the Charles Close Society,* 36 (April 1993), p. 15.

90 http://hansard.millbanksystems.com/commons/1955/nov/15/ordnance-survey-office-southampton.

91 http://www.newforest.gov.uk/index.cfm?articleid=4853&articleaction=dispmedia&mediaid=3027.

92 Thomas R. Tregear, *Hong Kong and the New Territories* (London: Hong Kong University Press, 1958), p. 69.

93 Lucien S. Vandenbroucke, 'The "Confessions" of Allan Dulles: New Evidence on the Bay of Pigs' in *Diplomatic History*, 8 (1984), pp. 377–80.

94 Peter Mangold, *The Almost Impossible Ally: Harold Macmillan and Charles de Gaulle* (London: I. B. Tauris, 2006), p. 166.

95 Peter Barberis, 'Introduction: The 1964 General Election – the "not quite, but" and "but only just" election' in *Contemporary British History*, 21:3 (September 2007), p. 286.

96 Figures based on the 1951 census stated that 44 per cent of all Glasgow housing was overcrowded.

97 The centre was to be cleared and re-planned (the Bruce Plan, 1946).

98 The 1956 Housing Subsidies Act (ended 1967) gave financial incentives for builders the higher they built.

99 David Lock, 'The Long View' in *Architectural Design Magazine: New Towns* (1994), p. 87.

100 Denis Cosgrove, *Apollo's Eye: A Cartographic Genealogy of the Earth in the Western Imagination* (Baltimore and London: Johns Hopkins, 2001), p. 258.

101 Op. cit., p. 263.

102 Robert Poole, *Earthrise: How Man First Saw the Earth* (New Haven and London: Yale, 2008), p. 5.

103 TNA, MINT 20/4254.

104 http://peterrmiles.wordpress.com/tag/bartholomews-maps/.

105 http://torquayfans.boards.net/thread/6771.

106 http://peterrmiles.wordpress.com/tag/bartholomews-maps/.

107 Matthew Taylor, *The Association Game: A History of British Football* (Harlow: Pearson Longman, 2008), p. 253.

108 Saville Report, 3.70.

109 The map explains that some areas are so sensitive as to have been omitted from the map.

110 This text was compiled from correspondence between the authors and Ron Jones, Jim Sharp and Stan McCaffrey (2012) and we are indebted to them for their insights into how the map was created. The replacement of the Strawberry Field gates with replicas was widely reported, for example in the *Guardian* of 10 May 2011; the campaign to save Madryn Street was also widely covered, for example in the *Daily Telegraph* of 14 June 2012.

111 Quoted from email correspondence of September 2012 (see note above).

112 Mahaffey explains the scale of the disaster by relating how radiation escape level warning alerts at a Swedish nuclear plant actually detected radiation from Chernobyl *entering* the facility. James Mahaffey, *Atomic Awakening* (New York: Pegasus, 2009), p. 318.

113 R. F. Pocock, *Nuclear Power: Its Development in the United Kingdom* (London: Institute of Nuclear Engineers, 1977), p. 226.

114 http://news.bbc.co.uk/onthisday/hi/dates/stories/november/3/newsid_2538000/2538155.stm.

115 At least one redrawing had been requested by Norway on the basis that the map was drawn on Mercator's projection and thus was in error of up to 1.5 degrees due to the curvature of the earth. See Alexander G. Kemp, *The Official History of North Sea Oil and Gas* (London: Routledge, 2012), vol. I, p. 65.

116 Op. cit., p. 68.

117 Peter D. Cameron, *Property Rights and Sovereign Rights: The Case of North Sea Oil* (London: Academic Press, 1983), p. 7.

118 All quotations are taken from *Silver Jubilee Beacons: An Informal Record of the National Network of Beacons to Celebrate the Silver Jubilee of Her Majesty Queen Elizabeth* (London: RICS, 1977).

119 Discussed fully by Alastair MacDonald in *Mapping the World: History of the Directorate of Overseas Surveys, 1946–85* (London: Stationery Office Books, 1996).

120 See Bob Parry and Chris Perkins, *World Mapping Today* (London: Bowker-Saur, 2000), p. 200.

121 Retrieved online from http://www.radiovop.com/index.php/national-news/7913-zim-s-largest-emerald-mine-shutdown.html.

122 Joe Moran, *On Roads* (London: Profile Books, 2010), p. 156.

123 Richard Cork, *Breaking Down the Barriers: Art in the 1990s* (New Haven: Yale University Press, 2003), p. 30.

124 John Barrow, *How Not to Make It in the Pop World: The Diary of an Almost Has Been* (Victoria, BC: Trafford Publishing, 2003), p. 26.

125 Eric Hobsbawm, *The Age of Extremes: The Short Twentieth Century* (London: Abacus, 1994), p. 280.

126 Edouard Chambost, *Using Tax Havens Successfully* (London: Institute for International Research, 1978), p. 21.

127 There is some confusion here, however, as the islands have been claimed by Argentina since 1925 and 1943 respectively.

128 John Nott, *Here Today, Gone Tomorrow* (London: Politico's, 2002).

129 'Cartography in Western Australia/Dept of Surveying and Cartography, Wembley Technical College, 1983 Technology and Further Education. Presented on the occasion of the 7th General Assembly and 12th Conference of ICA, Perth, Western Australia, 1984 (Western Australia State Library).'

130 John Pickles, *A History of Spaces: Cartographic Reason, Mapping and the Geo-Coded World* (London: Routledge, 2004), p. 33.

131 Peter Gill, *Famine and Foreigners: Ethiopia since Live Aid* (Oxford: Oxford University Press, 2010), p. 3.

132 J. A. Morris, quoted in Jeremy W. Crampton, 'Reflections on Arno Peters (1916–2002)' in *The Cartographic Journal*, 40:1 (June 2003), p. 55.

133 Michael Longan, 'Playing with landscape: social processes and spatial form in video games' in *Aether: The Journal of Media Geography*, vol. II (2008), pp. 23–40.

134 This text has been compiled from correspondence with William Wells and Stephen Humphries, April–December 2012.

135 Tony Judt, *Postwar: A History of Europe since 1945* (London: Heinemann, 2005), p. 707.

136 John Peters and John Nichol with William Pearson, *Tornado Down* (London: Michael Joseph, 1992).

137 John Davies, 'Uncle Joe knew where you lived: the story of Soviet mapping in Britain', part 1 in *Sheetlines: The Journal of the Charles Close Society* 72 (2005), pp. 26–38, and part 2 in *Sheetlines: The Journal of the Charles Close Society* 73 (2005), pp. 6–20. www.charlesclosesociety.org/files/Issue72page26.pdf and www.charlesclosesociety.org/files/Issue73page6.pdf. John M. Davies and Alexander J. Kent, *Soviet Intelligence Plans for the British Isles* (forthcoming).

138 Davies 2005, part 2, p. 26.

139 Cosgrove, 2001, *Apollo's Eye*, ibid.

140 Held in Atlanta, Georgia, in 1995, and published by the Institute of Electrical and Electronic Engineers later that year. The title of the paper, delivered by Stephen Eick of telecommunications company Bell-Labs, was '3D displays of the Internet'.

141 'Missile tracks across the net', http://mappa.mundi.net/maps/maps_008/.

142 Statistics from www.internetworldstats.com.

143 Martin Dodge and Rob Kitchin, *Atlas of Cyberspace* (London: Pearson Education, 2001).

144 Patrick W. Corrigan, 'Marlboro Man and the Stigma of Smoking' in *Smoke* (London: Reaktion, 2004), p. 347.

145 *Financial Times*, 26 January 2012.

146 Simon Garfield, *Off the Map: Why the World Looks the Way it Does* (London: Profile Books, 2012), p. 42.

147 Mark Monmonier, *Rhumb Lines and Map Wars: A Social History of the Mercator Projection* (Chicago: University of Chicago Press, 2004), p. 53.

148 www.famacollection.org.

149 www.famacollection.org/eng/fama-collection/fama-original-projects/17/index.html.

150 A great many 'parish maps' produced from the 1980s are listed on the Common Ground website.

151 One pioneering use of aerial photographs for historical research was the work of M. W. Beresford, who in *Medieval England: An Aerial Survey* (Cambridge: Cambridge University Press, 1958) analysed the medieval landscape through photographs.

152 Christopher Board, 'Air photo mosaics: a short-term solution to topographical map revision in Great Britain, 1944–1' in *Sheetlines: The Journal of the Charles Close Society*, 71 (December 2004), p. 28.

153 Martin Dodge, Chris Perkins and Rob Kitchin, *Rethinking Maps: New Frontiers in Cartographic Theory* (Abingdon and New York: Routledge, 2009), p. 228.

154 Laura Kurgan, 'Two projects with commentary: "Around Ground Zero" (map) and "New York, September 11, 2001 – four days later (satellite photographs)"' in Daniel J. Sherman and Terry Nardin (eds), *Terror, Culture, Politics: Rethinking 9/11* (Bloomington: Indiana University Press, 2006), p. 33.

List of Maps
All maps are lithographic prints on paper unless stated otherwise

A Trip to the Continent: Depicting a Trip from England to Berlin, via Paris. Board game, 46 x 72 cm. Private collection of Tim Bryars.

Imperial map of South Africa: Bloemfontein. Cape Town: Field Intelligence Dept., April 1900, 44 x 52 cm. Private collection of Tim Bryars.

The Navy League map, illustrating British naval history. Second edition, with corrections. Edinburgh and London: W & A.K. Johnston, 1901, 4 sheets, each 98 x 73 cm. British Library Maps 950.(138.).

George Arkell, Jewish East London in Charles Russell and H.S. Lewis, The Jew in London. A study of racial character and present-day conditions. London: T. Fisher Unwin, 1900. 40 x 60 cm. British Library 04034.ee.33.

Charles Booth, Life and Labour of the People in London ... Third Series: Religious Influences. Volume 2, London North of the Thames: The Inner Ring. London: Macmillan, 1902. Facing p. 111, 36 x 28 cm. Private collection of Tim Bryars.

H. Delmé-Radcliffe, Military sketch map to show Sir R. Buller's advance from Chieveley to relieve Ladysmith, Feby. 14th to March 1st, 1900. London: E. Stanford, 1902. 54 x 50 cm. British Library Maps RUSI V.17.5A

Cardiff, Cogan, Penarth and Taffs Well from Railway Junction Diagrams. London: Railway Clearing House, Euston Square, 1905. Engraved by J. P. & W. R. Emslie. 17 x 27 cm. British Library Maps C.44.d.86.

James Montgomery Flagg, A Map of the World as Seen by Him. London: James Henderson & Sons, [c. 1907]. Postcard, 14 x 9 cm. Private collection of Tim Bryars.

Opium, Drugs, &c. from J. Bartholomew (ed.), Atlas of the World's Commerce. London: G. Newnes, [1907]. 36 x 44 cm. British Library Maps 48.e.9.

Asia, from Philips' Series of Model Test Maps. Set 2. London: George Philip & Son, [1908]. 85 x 68 cm. British Library Maps 938.(2.).

"How To Get There" An Interesting and Educational Game for 2, 3, or 4 Players. London: Johnson, Riddle & Co. Ltd., [c. 1908]. 28 x 36 cm. British Library Maps 188.v.32.

Medicine Hat, Alberta, by C. E. Goad. London and Montreal, 1910. 53 x 63 cm. British Library Maps 146.b.48.(25.).

Ulster's Prayer Dont Let Go! Belfast: John Cleland & Sons, c. 1911. Postcard, 9 x 14 cm. British Library Maps C.1.a.9.(357.).

Coronation Durbar Delhi 1911. London: Army & Navy Co-operation Society, 1911. Lithograph on silk, 52 x 41 cm. Private collection of Tim Bryars.

"Hark! Hark! The Dogs Do Bark!" With note by Walter Emanuel. Designed ... by Johnson, Riddle & Co. [London]: G. W. Bacon & Co., [1914]. 71 x 49 cm. British Library Maps 1078.(42.).

Ernest Shackleton, [A sketch map of Antarctica]. London, 1914. Pencil on paper, 18 x 24 cm. Royal Geographical Society, EHS/1/a)1914

B. Crétée, [Europe in 1914-1915]. Warsaw: Vladislav Levinsky, 1915. 43 x 56 cm. Private collection of Tim Bryars.

Map of the area occupied by Australian and N.Z. corps [Imbros]: G.H.Q. M.E.F., 1915. Lithograph with pencil annotations, 55 x 38 cm. British Library Maps 43336.(21.).

The German Scheme of Mittel Europa. London: Sifton, Praed & Co. [c. 1916]. 251 x 183 cm. Courtesy of Sifton, Praed / The Map House

Trench Map, Montauban, scale 1;20 000. Southampton: Ordnance Survey, 1916. Sheet 37 x 63 cm. British Library Maps RUSI B.587.

Bernard Sleigh, An Ancient Mappe of Fairyland Newly Discovered and Set Forth. London: Sidgwick & Jackson, [1918]. 48 x 182 cm. British Library L.R.270.a.46.

Map of Amritsar city, in Report of the committee appointed by the government of India to investigate the disturbances in the Punjab, &c. London: H. M Stationery Office, 1920. 30 x 33 cm. Private collection of Tim Bryars.

[A Map of the New Greater Greece]. London: Hesperia Press, 1920. Sheet 70 x 50 cm. Private collection of Tim Bryars.

Atlas of the British Empire. Reproduced from the Original made for Her Majesty Queen Mary's Doll's House. London: E. Stanford Ltd., [1924]. Atlas, 4 x 5 cm. British Library Maps C.7.a.26.

The County of London, GSGS. no. 3786A. sheet 1/sheet 2. Southampton: Ordnance Survey Office, 1926. 2 sheets, each 100 x 75 cm. British Library Maps CC.5.a.170.

Ordnance Survey Map of the Solar Eclipse, 29th June, 1927. Southampton: Ordnance Survey Office, 1927. 60 x 83 cm. British Library Maps 1135.(12.).

Ernest H. Shepard, Hundred Acre Wood in A. A. Milne, Winnie-the-Pooh. London: Methuen, 1926, 9 x 25 cm. British Library C.98.k.15.

A Pictorial Chart of English Literature, Compiled by Ethel Earle Wylie. Illustrated by Ella Wall Van Leer. Chicago: Rand McNally, 1929. 66 x 87 cm. British Library Maps 1080.(78.).

The Heart of Scotland. Gleneagles. Constructed and printed by W. & A. K. Johnston. Glasgow: McCorquodale & Co., [c. 1930]. 104 x 72 cm. British Library Maps 1.c.54.

Palestine. Map no. 6. GSGS 3743. Jaffa: Survey of Palestine, in Sir John Hope Simpson, Report on Immigration, Land Settlement and Development in Palestine, London: HMSO, 1930. 118 x 56 cm. Private collection of Tim Bryars.

MacDonald Gill, A Map of New Zealand, Portraying Her Agricultural Products & Fisheries. [London]: Empire Marketing Board, [1930]. 147 x 97 cm. British Library Maps 92501.(9.).

Sidney 'George' Strube, 'A Chart of the financial main: here lie millions in sunken gold', published in Daily Express, 23rd January 1933.

Plan von Nürnberg mit Aufmarschstrassen der Festzüge. München & Berlin: Zentralverlag der NSDAP, 1936, sheet 33 x 41 cm. Private collection of Tim Bryars.

Coronation - King Edward VIII - 1937. London: "British made", [1936]. Print on fabric, 23 x 33 cm. British Library Maps CC.5.a.298.

Mapa Politico España, autorizado por la Delegación Provincial de Propaganda. Bilbao: Centro Grafico Iris, [1938]. 61 x 75 cm. British Library Maps X.1956.

London Mayfair Square / Neues Luftfahrtministerium Techn. Amt. [Berlin: Luftwaffe, 1940]. 30 x 31 cm. British Library Maps CC.5.a.555.

London County Council, London sheet v.9. Southampton: Ordnance Survey, 1916 (1934) (with hand colour annotations to 1945). Map no. 61. 61 x 102 cm. London Metropolitan Archives, LCC/AR/TP/P/38-43.

Pariser Plan. Paris: Mouillier & Dermont, Oktober 1940. 37 x 52 cm. Private collection of Tim Bryars.

[A Japanese map of Pearl Harbor] 1941. Ink on paper, dimensions unknown. Robert L. Lawson photograph collection, U.S. Navy National Museum of Naval Aviation, photo no. 1996.488.029.049.

'S.P.K.', Confiance ... Ses amputations se poursuivent méthodiquement. [Paris?, c. 1942]. 119 x 84 cm. British Library Maps CC.6.a.39.

Frederick Donald Blake, [The Battle of the Atlantic. London?, c. 1943]. 71 x 46 cm. Private collection of Tim Bryars.

New York City Transit System. New York: George J. Nostrand, [c. 1944]. 41 x 23 cm. Private collection of Tim Bryars.

Supreme Headquarters Allied Expeditionary Force. Enemy Defences, ALDERNEY. [London]: War Office, 1944. 47 x 59 cm. British Library Maps MOD GSGS 2558.

Caen. GSGS 4347 2nd edition. [London]: War Office, 1944. Sheet 50 x 62 cm. Private collection of Tim Bryars.

Occupation of Women by Regions, 1931 – Chart IV, from Gertrude Williams, Women and Work. London: Nicholson & Watson, 1945. 21 x 14 cm. Private collection of Tim Bryars.

Heinz Schunke, Neue Wege durch Berlin ... Berlin: M. & R. Meier, September 1945. 15 x 21 cm. Private collection of Tim Bryars.

Greetings Nürnberg Germany 1945. [Nürnberg? n.p.] 17 x 9 cm. Private collection of Tim Bryars.

Boundary Commission Award of Bengal. Calcutta: Dipti Printing Co., 1947. 70 x 45 cm. Private collection of Tim Bryars.

Estra Clark, A Map of Yorkshire Produced by British Railways. London: Published by the Railway Executive (Eastern & North Eastern Regions), 1949. 87 x 108 cm. British Library Maps CC.6.a.35.

Reginald Montague Lander, Explore the Yorkshire Coast and Nearby Countryside by Train. British Railways (North Eastern region, 1963. 102 x 127 cm. National Railway Museum, 1978–9129.

Francis Chichester, Aquila Airways. [London, c. 1949]. 44 x 34 cm. Private collection of Tim Bryars.

Geographical Magazine, What do they talk about? London: E. G. R. Taylor and C. W. Bacon, 1951. 66 x 44 cm. Private collection of Tim Bryars.

Brian Orchard Lisle, Map of Caribbean Oil .... Fort Worth, 1952. 124 x 95 cm. British Library Maps 83001.(44.).

Royal Tour 1953–54: Outward Journey / Homeward Journey. London: Prepared by the Hydrographic

Dept. of the Admiralty, 1953. 2 maps, each 70 x 103 cm. British Library Maps C.49.e.64.

'D. E. B.', Cunard Caronia Coronation Cruise. 'Printed in England', n.p. 1953. 54 x 65 cm. Private collection of Tim Bryars.

[A Map of Middle Earth] published in J. R. R. Tolkien, The Lord of the Rings. London: George Allen & Unwin, 1954. 41 x 41 cm. British Library NN.7521.

Suez Canal Zone Special Series [Survey Directorate, Middle East], 1954. 71 x 56 cm. British Library Maps MOD MDR Misc 11922 sheet 6.

GSGS 4677 Sheet SU41. Chessington: Ordnance Survey, 1957. Sheet 56 x 52 cm. British Library Maps MOD GSGS 4627.

A Nuñez & Javier Soler, Alicante. [Alicante]: Junta Provincial del Turismo, 1957. 45 x 64 cm. British Library Maps Y.1612.

The World We Live In. Compiled for the Barclays Group of Banks by David L. Linton ... Designed and drawn by E. W. Fenton. Ipswich: W. S. Cowell, [1958]. 49 x 96 cm. British Library Maps 37.b.55.

Thomas R. Tregear, Hong Kong And The New Territories/ GSGS 3961. War Office / Hong Kong University Press, 1958. 64 x 88 cm. Private collection of Tim Bryars.

Cuba 1:250,000, SERIE ICCC E522. Sheet NF 17-6 [Havana]: Instituto Cubano de Geodesia y Cartografia, 1961. 56 x 70 cm. British Library Maps Y.2601.

Industry I: Selected industries in The European Community ... Prepared by Dr. I. B. F. Kormoss ... with the advice of M. Gabriel Quencez. Brussels & Luxembourg: Press and Information Service of the European Communities, 1962. 12 maps, each 20 x 23 cm. British Library Maps 35.c.40.

The Times Map for the General Election, 1964. London: Times Publishing Co., 1964. Lithograph with pencil annotations, 49 x 55 cm. British Library Maps 1092.(27.).

Glasgow: Residential Land Use 1965/designed and compiled by Michael Wood from an original survey and classification by D. R. Diamond. Glasgow: Department of Geography, University of Glasgow, 1968. 33 x 42 cm. British Library Maps Y.1260.

New Towns Act 1965. The North Buckinghamshire (Milton Keynes) New Town (Designation) Order 1967. Southampton: Ordnance Survey for the Ministry of Housing and Local Government, 1967. 70 x 73 cm. British Library Maps 1190.(217.).

William Anders, 'Earthrise', 1968. Colour photograph. NASA, AS8-14-2383.

[Royal Geographical Society medal commemorating the first landing on the moon] London, 1969. Silver medal, 6.5 cm. diameter. British Library Maps R.M.17.

Football History Map of England and Wales Approved by the Football Association. Compiled by John Carvosso. Edinburgh: John Bartholomew & Son, [1971]. 98 x 74 cm. British Library Maps 1190.(177.).

Richard Long, A Hundred Mile Walk. 1971–2. Graphite on map, typescript, photograph and printed labels on board, 22 x 49 cm. Tate T01720.

Paddy Allen and Paul Scruton, Bloody Sunday: Interactive Map. theguardian.com, Tuesday 15th June 2010.

http://www.guardian.co.uk/uk/interactive/2010/jun/ 10/northern-ireland-bloody-sunday-interactive-map

Map of Cherished Land, published in The Geographical Magazine. London: IPC Magazines, October 1973. 51 x 76 cm. British Library Maps 1080.(112.).

The Areas for Expansion: Extensive Government Aid for Industry in Special Development Areas, Development Areas, Intermediate Areas, Northern Ireland. [London]: Department of Industry, [1977]. 99 x 72 cm. British Library Maps 1082.(13.).

Beatles Map, from The Beatles Collection: from Liverpool to the World. City of Liverpool Public Relations Office 1974, 46 x 41 cm. Private collection of Tim Bryars.

Nuclear Engineering International, Map of the World's Nuclear Power Plants. London: IPC Business Press Ltd., 1975. 54 x 90 cm. British Library Maps X.4196.

Nordsjøoljen leteaktiviteter og funn. ESSO, 1975. 69 x 43 cm. British Library Maps X.4825.

Geoffrey Taylor, San Serriffe. London, 1977. Ink and typescript on tracing paper, 26 x 21 cm. British Library Maps CC.5.a.429.

Fairey Surveys Ltd. Silver Jubilee Beacons 1952–1977. [London]: Royal Institution of Chartered Surveyors, [1977]. 81 x 55 cm. British Library Maps X.681.

Zeus Mine, Rhodesia. Salisbury: Surveyor General, 1978. 85 x 61 cm. Private collection of Tim Bryars.

Happy Eater Family Restaurants Route Planner, London? c. 1980, 30 x 21 cm. Private collection of Tim Bryars.

Whitbread Brewery Co., 15 Inns of Character in Kent. London: Presswork Publications Ltd., [c. 1981]. 51 x 42 cm. Private collection of Tim Bryars.

The Islands of the Blessed in Michael Kidron & Ronald Segal, The State of the World Atlas: A Pluto Press project. London: Pan, 1981. 22 x 33 cm. British Library Maps 60.b.39.

'FAGA', British Antarctic Territory, postcard no. 352 (revised state). Southampton: Southern Printers, [c. 1982]. Postcard, 15 x 16 cm. Private collection of Tim Bryars.

London Gay City Map: A Spartacus Production. Edinburgh: John Bartholomew & Son Ltd., 1982. 74 x 99 cm. British Library Maps 212.f.30.

David Llewellyn, 'The World of George Orwell's "1984"'. Wembley, Australia, 1983. Pen and ink on paper, 28 x 40 cm. British Library Maps X.680.

CND. Cruise Deployment: roads, telephones, bases. [Southampton]: R. Burnell, Inner City branch, Southampton CND, 1984. 40 x 28 cm. British Library Maps CC.5.a.416.

[Live Aid Logo, 1985]. Print on fabric, dimensions unknown.

Harrier & Beagle Hunts of Great Britain Including Kennels. London: Hampton Editions, 1986. 57 x 39 cm. British Library Maps X.506.

Institute of Grocery Distribution. Superstores 1986. Watford, 1986. 80 x 53 cm. British Library Maps X.1536.

Screengrabs from SimCity. Maxis Productions, 1989. Video game.

Two laminated maps of Belfast, cut down from the Ordnance Survey and overprinted. c. 1990. 28 x 21 cm or smaller. Private collection of Tim Bryars.

ESCAPE AND EVASION MAPs 1:1,000,000, GSGS. 5619. London: Director General Military Survey Ministry of Defence, 1990. Print on Paxar. sheet 46 x 90 cm. British Library Maps X.4578. Screengrab from CNN News, Broadcast 1990.

Sanderlend. [Moscow?]: Generalńyĭ Shtab, 1976. 125 x 92 cm. British Library Maps X.4614.

Orlando including Walt Disney World. Chicago: Rand McNally & Company, [1994]. Folded to 32 cm. British Library Maps 220.a.737.

K. C. Cox & S. G. Eick, [Still taken from a computer simulation], '3-D Displays of Internet Traffic' in Proceedings of the '95 Information Visualization Conference, Atlanta, GA. Los Alamitos, CA: Institute of Electrical and Electronic Engineers Computer Society Press, 1995. 28 x 22 cm. British Library 4496.403000.

Philip Morris Co., Where will they draw the line? 2 adverts, published in The Economist, 17 June 1995 and 13 July 1995. Each 27 x 21 cm. British Library Maps 187.v.1.

'Ethnic Minority Residents 1991', in Danny Dorling, A New Social Atlas of Britain. Chichester: Wiley, 1995. 30 x 22 cm. British Library Maps 194.b.40. http://sasi. group.shef.ac.uk/publications/new_social_atlas/

Gerard Mercator [The Mercator Atlas of Europe. Duisburg, 1570–72.]. 46 copper engraved and manuscript maps, 40 cm. British Library Maps C.29.c.13.

Where Are You Now? Your Map of the Glastonbury Festival 1997. Published in The Official Programme of the 1997 Glastonbury Performing Arts Festival. London: Select Magazine, 1997. 30 x 41 cm. Author's collection.

[A map showing the route of Diana's funeral procession] Published in Daily Mirror, September 2nd 1997, p. 3.

Suada Kapic and Miran Norderland. The Fall of Yugoslavia, 1991–1999. [Sarajevo]: FAMA International, 1999. 81 x 110 cm. British Library Maps X.4982.

Enid Pepper and others, 'Sherington'. Sherington, 1999– 2004. Embroidery, 190 x 200 cm. Sherington Historical Society.

VIZ, Cuntinental Europe. London: John Brown, 2001. 80 x 56 cm.

Getmapping.com. [An aerial photographic view of Coventry] in England: The Photographic Atlas. London: HarperCollins, 2001, pp. 458-9. 45 x 61 cm. British Library Maps Ref F.6.(Eur) (The) 2.

[Aerial photo-mosaic of Coventry]. 42/37 N.W. Chessington: Ordnance Survey, 1949-53. 54 x 51 cm. British Library Maps O.S.M.

Laura Kurgan. Around Ground Zero, Dec. 2001. New York: New York New Visions, [2001?]. 61 x 46 cm. British Library Maps X. 6952.

Further Reading

James R. Ackerman and Robert W. Karrow Jr (eds),
Maps: Finding Our Place in the World (Chicago:
University of Chicago Press, 2007)

Peter Barber (ed.), The Map Book
(London: Weidenfeld & Nicolson, 2005)

Jeremy Black, Maps and Politics
(London: Reaktion, 1997)

Piers Brendon, The Decline and Fall of the British Empire
(London: Jonathan Cape, 2007)

Jerry Brotton, A History of the World in Twelve Maps
(London: Penguin, 2013)

Karen Cook, 'The historical role of photomechanical
techniques in map production' in
Cartography and Geographic Information Science,
29:3 (2002)

Denis Cosgrove, Apollo's Eye: A Cartographic Genealogy
of the Earth in the Western Imagination
(Baltimore: Johns Hopkins University Press, 2001)

Jeremy W. Crampton, 'Exploring the history of
cartography in the twentieth century' in
Imago Mundi 56:2 (2004)

John Darwin, The Empire Project: The Rise and Fall
of the British World System, 1830-1970
(Cambridge: Cambridge University Press, 2009)

Catherine Delano-Smith and Roger J.P. Kain,
English Maps: A History
(London: British Library, 1999)

Martin Dodge, Chris Perkins and Rob Kitchin,
Rethinking Maps: New Frontiers in Cartographic
Theory (Abingdon: Routledge, 2009)

Felix Driver & David Gilbert (eds),
Imperial Cities: Landscape, Display and Identity
(Manchester: Manchester University Press, 1999)

David Forrest, 'The top ten maps of the twentieth
century' in The Cartographic Journal 40:1
(June, 2003)

Simon Garfield, On the Map: Why the World Looks
the Way it Does (London: Profile, 2012)

J.B. Harley, Ordnance Survey Maps: A Descriptive Manual
(Southampton: Ordnance Survey, 1975)

J.B. Harley and David Woodward (eds.), The History of
Cartography Volume One: Cartography in Prehistoric,
Ancient and Medieval Europe and the Mediterranean
(Chicago and London: University of Chicago Press,
1987)

J.B. Harley (Paul Laxton ed.), The New Nature of Maps:
Essays in the History of Cartography
(Baltimore: Johns Hopkins University press, 2001)

P.D.A. Harvey, The History of Topographical Maps:
Symbols, Pictures and Surveys
(London: Thames & Hudson, 1980)

Michael Heffernan, 'The cartography of the fourth
estate: mapping the new imperialism in British and
French newspapers, 1875-1925' in James R. Ackerman

(ed.), The Imperial Map: Cartography and the Mastery
of Empire (Chicago: University of Chicago Press, 2009)

Eric Hobsbawm, Age of Extremes: The Short Twentieth
Century 1914–1991 (London: Abacus, 1994)

Tony Judt, Postwar: A History of Europe since 1945
(London: Heinemann, 2005)

Geoff King, Mapping Reality: An Exploration of Cultural
Geographies (Basingstoke: Macmillan, 1996)

Keith Lowe, Savage Continent: Europe in the Aftermath
of World War II (London: Viking, 2012)

Maria Misra, Vishnu's Crowded Temple: India since the
Great Rebellion (London: Allen Lane, 2007)

Mark Monmonier & David Woodward, 'The exploratory
essays initiative: background and overview' in
Cartography and Geographic Information Science,
29:3 (2002)

Mark Monmonier (ed.), The History of Cartography
Volume Six: Cartography in the Twentieth Century
(Chicago and London: University of Chicago Press,
forthcoming)

C.R. Perkins and R.B. Parry, Mapping in the UK
(London: Bowker-Saur, 1996)

John Pickles, A History of Spaces: Cartographic Reason,
Mapping and the Geo-coded World
(London: Routledge, 2004)

Norman Stone, World War One: A Short History
(London: Allen Lane, 2007)

Picture Credits

pp. 44–45 Royal Geographical Society; pp. 52–3 image courtesy of Sifton Praed/The Map House; p. 67 Royal Collection Trust/© Her Majesty Queen Elizabeth II 2014; p. 71 copyright © The Estate E.H. Shepard Trust, reproduced with permission of Curtis Brown Limited, London; p. 82 Photo British Cartoon Archive, University of Kent, www.cartoons.ac.uk, © Daily Express; pp. 93–94 London Metropolitan Archives; p. 98 US Navy National Museum of Naval Aviation, Penascola; p. 121 © NRM/ Pictorial Collection/Science & Society Picture Library; p. 133 © Estate of J.R.R. Tolkein; pp. 158–59 Photo Tate, London 2014 © DACS, 2014; p. 161 top & centre Copyright Guardian News & Media Ltd. 2010; p. 161 below Rex Features; pp. 192–93 Michael Ochs Archive/Getty Images; p. 198 Sim City images used with permission of Electronic Arts Inc.; p. 202 Getty Images; pp. 202-03 © UK MOD Crown Copyright, 2014; p. 217 Mirrorpix; p. 221 © FAMA Collection, 1999 www.famacollection.org; p. 224 Simon Thorpe/Viz.

Index Figures in *italic* refer to pages on which illustrations appear.